ROUTLEDGE LIBRARY EDTIONS: GLOBAL TRANSPORT PLANNING

Volume 11

TRANSPORT AND PUBLIC POLICY

TRANSPORT AND PUBLIC POLICY

K.M. GWILLIAM

Routledge
Taylor & Francis Group

LONDON AND NEW YORK

First published in 1964 by George Allen & Unwin Ltd

This edition first published in 2021
by Routledge
2 Park Square, Milton Park, Abingdon, Oxon OX14 4RN

and by Routledge
605 Third Avenue, New York, NY 10017

Routledge is an imprint of the Taylor & Francis Group, an informa business

© 1964 George Allen & Unwin Ltd

British Library Cataloguing in Publication Data
A catalogue record for this book is available from the British Library

ISBN: 978-0-367-69870-6 (Set)
ISBN: 978-1-00-316032-8 (Set) (ebk)
ISBN: 978-0-367-74057-3 (Volume 11) (hbk)
ISBN: 978-0-36774-077-1 (Volume 11) (pbk)
ISBN: 978-1-00-315599-7 (Volume 11) (ebk)

Publisher's Note
The publisher has gone to great lengths to ensure the quality of this reprint but points out that some imperfections in the original copies may be apparent.

Disclaimer
The publisher has made every effort to trace copyright holders and would welcome correspondence from those they have been unable to trace.

Transport and Public Policy

BY

K. M. GWILLIAM

London
GEORGE ALLEN & UNWIN LTD
Ruskin House   Museum Street

PRINTED IN GREAT BRITAIN

in 10 on 11 point Times Roman type

BY SIMSON SHAND LTD
LONDON, HERTFORD AND HARLOW

PREFACE

In lecturing on the economics of transport to students at Nottingham University I have, for some time, felt the lack of an adequate basic text. No disrespect is thereby meant to previous writers in this field. It is simply that the transport world, and particularly public policy for transport, has changed so rapidly and radically in the last few years that books of less than a decade ago already seem archaic. Mr Peter Donaldson, teaching in the same field at Leicester University, was of a like mind and suggested that we should together produce a book to fill this gap. Unfortunately, soon after beginning the work, his departure to take up an appointment in India made continued collaboration impracticable. Nevertheless, apart from providing the original stimulus, Mr Donaldson has had considerable influence on the general form of the book and in more detail on the content of Chapters 2, 3 and 10. I am very grateful for his contribution to the enterprise.

Many others have subsequently given encouragement and help. I would particularly thank Professor F. A. Wells, Messrs A. Gabor, A. J. Harrison, J. A. B. Hibbs, C. K. Rowley and J. W. Smith for most valuable comments and suggestions. Above all I would like to thank my wife both for her passive endurance of lost attention during the last year and for her active encouragement and support in completing the work.

K. M. G.

Nottingham
December 31st, 1963

CONTENTS

INTRODUCTION

It is not difficult to find justification for writing a book about transport. As individuals we are constantly aware of transport problems; our train may be late, our car may get caught in a traffic jam or we may be concerned by what appears to be an abnormally high bus fare. Indeed, the transport system must rank with the weather amongst the most frequent butts for our grumbles and our jokes. Perhaps it provides an even more pleasing target than the weather for we are dealing in this case not with an accident of nature but an artifice of man, the defects of which can be attributed to some person or body. This is particularly satisfying when that body happens to be some kind of public authority which can be criticized and maligned with impunity. Few critics, however, are well informed on the exact location and division of responsibility for the transport system; even fewer really appreciate the nature of the problems involved in devising a rational transport system. This book is an attempt to stimulate such appreciation and, in this light, to assess the role of the government and its agencies in the transport sector. Though it is primarily intended for the university economics student it is also hoped that the layman seeking an informed and ordered view of the transport sector will not experience undue difficulty in following the argument.

In Part I we attempt to set up a framework of accepted economic principles concerning the efficient operation of a transport system. This task is in itself difficult for economists, like the Church, rarely speak with a single voice when they are asked to extend their analysis beyond the most simple and tautological principles. Despite, or perhaps because of, the complications involved in this theoretical background it is essential to our consideration of the current situation in the transport sector. This is because it is from economic theory that we derive the main thesis of the book, namely, that basically there is one central problem of transport co-ordination. Unless we acknowledge the essential connection between the apparently separate problems of various forms of transport and realize that they must be considered simultaneously, we could embrace public policies towards transport which may in the long run be regretted. In some American cities, for example, attempts arc now being made, at tremendous cost, to resuscitate a public transport system previously destroyed by the extension of private motoring.[1]

[1] See John S. Gallagher, Jr., 'Urban Transport Requirements in the U.S.A.', *British Transport Review*, Vol. 6, No. 2, December 1956.

In Part II we trace the history of Government intervention in transport. We would justify the inclusion of this section on three grounds. Firstly, few industries have been so bound by ancient restrictions. For instance, the obligation not to show 'undue preference', which limited railway charging until 1953, was almost a hundred years old and was based on a market situation that had disappeared long ago. In many respects it is impossible to understand the apparent illogicalities of the structure unless one has some appreciation of its historical antecedents. Secondly, we wish to show that public policy for transport has not always been concerned with the same problems; changed conditions may call for revision in the attitudes of public authorities. Thirdly, we may note the similarities as well as the differences between historical situations. Changes of technology cause shifts in the balance of advantage between agencies. In a private enterprise system the declining sectors naturally tend to resort to defensive strategies which would retard progress. As canals superseded roads at the beginning of the nineteenth century and railways superseded canals in the middle of the century, so it now appears that road transport is coming into its own at the expense of the railways. We may note from our historical analogy the danger that public policy may be misguided if it merely protects the interests of existing operators against more progressive elements.

In Part III we consider the current organization of the industry. The individual chapters, though intended to be complete in themselves as descriptions of particular agencies, also indicate the extent of complementarity or competition between agencies.

Finally, in Part IV, we take up the question of these relationships in more detail by considering the problem of transport co-ordination. We begin by discussing the theoretical merits of some suggested, but untried, means of co-ordination, continue by critically examining the approach currently applied under the legislation of the early 'sixties and wind up, in Chapter 14, by investigating some of the possible implications for transport co-ordination of British entry into the European Economic Community.

We argue throughout that transport presents a single problem in the sense that public control of one agency inevitably has repercussions on others and should therefore only be undertaken after full consideration of the effects on the whole transport sector. We do not, however, claim that it is a single problem in the sense that the resolution of a single conflict will dispose of it entirely. There are at least three conflicts which are, from time to time, mistakenly regarded as the sole element in the problem. The first, concerning the form of transport, is primarily a conflict between road and rail interests. The second is that between the private carrier (the 'C' licence haulier, the contract coach operator and the private motorist) and the public carrier (the 'A' licence haulier, the railways and the licensed bus service operator). The third is the political conflict concerning ownership between the supporters of nationalization and those of private enterprise. These conflicts are separate both in principle and in practice. For instance, though the 1947 Transport Act provided for the nationalization of most forms of public transport it did not eliminate competition between road and rail or between private and public transport.

Thus our contention that the problem is single does not entail that it is simple; indeed it is the very complexity of the problem which makes it so intractable. We do not pretend in this book to have found a simple key which will eliminate the need for detailed consideration of all the complexities of the transport sector. On the contrary we suggest that the solution lies in ensuring a basis of fair comparability between competing sectors which can only be obtained by detailed and comprehensive investigation of the industry. It follows from this general approach to the subject that much of the book will be concerned with compilation of the relevant information. But facts without interpretation are as barren as faith without works. It is hoped that our analysis will also stimulate the reader to develop sensible and informed opinions in a field where misunderstanding and lack of adequate information are so common.

PART I

ECONOMIC PRINCIPLES AND THE TRANSPORT SECTOR

CHAPTER ONE

The Scope of Public Policy

It is impossible to define the scope of public policy for transport both precisely and objectively. The exact role of the State, in transport as in other sectors, is a matter for political judgment. It is possible though, by retaining a high level of generality, to produce a statement of the aims of public policy which is sufficiently meaningful to provide a useful starting point for our study without being politically contentious.

THE ALLOCATION OF RESOURCES BETWEEN SECTORS

Even if we made the simplest possible assumption concerning the product of our industry, namely that transport services constitute a single homogeneous product, we would still need to ensure that there was no overall misallocation of resources as between transport and other sectors of the economy.

We might define this desirable allocation as that which maximized the net social product. Or, to put it another way, as a position where the net social productivity of a marginal increment of resources was equal in all sectors. Unfortunately, neither the concept of net social product nor that of marginal social productivity can be usefully measured in such a way as to give the statement empirical meaning. The problem of the identification and valuation of the social product, only part of which arises in such a way as to have a value assigned to it by a market, would defeat us in both cases.

In vivid contrast to such an abstractly theoretical approach we might attempt to make geographical or temporal comparisons of the proportion of national resources devoted to transport. Certainly the proportion of the total labour force employed in transport and communications varies widely, from 2 per cent in India and Pakistan to 7·7 per cent in U.K. and 9·2 per cent in Australia.[1] But the demand for transport services depends on such diverse considerations as the terrain, the industrial structure and the trading relationships, all of which vary from country to country. Direct international comparisons are therefore of no particular significance. Nor are we any better placed when we consider the U.K. figure over a period of time. The

[1] See *United Nations Yearbook of Labour Statistics*, 1962. Table 4.

B

slight decline in the actual proportion of the working population employed in transport does not, by itself, give any indication of what the proportion ought to be.[1]

Perhaps a closer investigation of the transport sector itself may yield useful conclusions. For instance, if there is evidence that output has been lost as a result of the inability of the transport sector to meet the demands made of it, we might argue for more resources to be devoted to transport. Or, on the other hand, if there is evidence of waste of resources or low productivity in transport operation we might argue that existing demand could be met with less call on scarce resources.

The freedom to provide transport on own account makes it possible for any trader who is dissatisfied with the service provided by public operators to meet his own requirements. In this way the proportion of national resources devoted to transport can always be increased. Thus it might be argued that the free operation of the market mechanism will ensure the devotion of adequate resources to transport, in the quantitative sense at least. There are, however, some forms of investment beyond the competence of even the largest firm. Road congestion reduces the operating efficiency of road haulage operators, both private and public. It could be eliminated, or at least reduced, if enough resources were devoted to dealing with the problem. As no individual firm can be expected to make the huge investments required to attack the problem the matter devolves upon public authorities. It then becomes necessary to decide whether the advantages to be gained from road improvements by individuals and firms are large enough to justify the cost to the public authority of making them. Traditionally the price mechanism has not been applied in this field in Britain with the result that public authorities have been unable to subject their investments in roads to the normal commercial test of profitability. But decisions have to be made. Unless public investment is to be undertaken in a random and unscientific manner public authorities must find some criterion for making them. Thus the development of a criterion for road investment is a necessary step towards establishing an economic allocation of resources between sectors.

A slightly different problem arises in the case of railways. Closure of branch lines and the withdrawal of services, so long as they do not involve the substitution of other facilities more costly to the community, represent a net reduction of the quantity of resources devoted

---

[1] The argument has been simplified here by considering only labour requirements. Further problems of comparison and aggregation would arise if other factors were also taken into account.

to transport. Again, however, a question of valuation arises for the value of the resources saved has to be weighed against the value of the service withdrawn. Current policy in this respect is based on the judgment that a service is not worth retaining unless, in aggregate, those using it are willing to meet the full costs of providing it. Nevertheless, claims for special consideration are common and the treatment of such claims is clearly germane to the issue of the distribution of resources between sectors. Before we can hope to deal sensibly with exceptions, however, we must first understand the reasoning behind the general rule that services should only be provided in response to adequate effective demand, i.e. demand backed by a willingness to pay enough to meet full costs.

The role and nature of demand for transport
The real cost of consuming any commodity or service is the most desirable alternative foregone as a result of the decision to consume it. It follows, therefore, that the individual will be getting the most satisfaction from his income only if he values each unit of consumption more highly than its cost, that is, more highly than the alternatives foregone. To the consumer, who has some appreciation of the money prices of other goods, the money price of a particular commodity or service is an indication of the real alternatives he will forego by purchasing it. Thus, by producing a service for which consumers are not willing to pay the full cost, we are, by the consumers' own indication, failing to get the best out of the resources so used.

Now it is generally the case that the more one already has of a particular commodity the less will be the value of any subsequent additional units. To put it into economic jargon, a condition of diminishing marginal utility generally obtains. In seeking to get the most satisfaction from his income the individual will extend his consumption of each commodity until the additional value of the marginal unit ceases to be greater than the price. The higher the price the sooner will this point be reached. It follows also that any increase in price will normally cause less to be purchased and conversely any reduction in price will cause more to be bought. If a price reduction induces such a large increase in sales that total expenditure on the commodity in question increases we say that demand is 'elastic' with respect to price. If, on the other hand, a price reduction is associated with a reduction in expenditure we describe demand as 'inelastic'.

There are two determinants of elasticity of demand which have particularly important implications in transport, namely, 'substitutability' and 'time'.

The closer are the available substitutes the more elastic will the

demand for the service tend to be. But there is greater substitutability between different forms of transport than between transport
and other commodities or services. Therefore, though a change in the
relative prices of road and rail transport may cause considerable
redistribution of traffic between the agencies, a general rise in the
price of transport may not immediately cause any great reduction in
total demand.

But we must also consider the time element. If a reduction in
price is maintained long enough it will eventually affect the location
of industry. The incentive to minimize transport requirements will
be weakened and the demand for transport may therefore increase in
physical terms as a result of the locations chosen for newly established
plant. Once a plant is established it may become very expensive to
move it again. So long as the price of transport represents a main-
tainably low level of cost no harm is done. If it does not, then, because of the great costs of relocation, the economy may for a long
time be committed to a pattern of industrial location involving higher
total costs than would be the case if an alternative pattern of location prevailed.

We may summarize the position as follows. Though in the short
run the demand for transport as a whole tends to be relatively inelastic and prices function only to determine the relative importance
of different forms of transport, in the longer run the location of
industry will be affected by the general level of transport prices. Thus,
whatever the structure of the transport system, it is important, both
from the point of view of the immediate allocation of traffic between
agencies and from that of the long-term development of the economy,
that the prices charged should be related to the costs incurred. The
difficulties involved in giving this suggestion an operative meaning
are discussed in Chapter 3.

THE EFFICIENCY OF TRANSPORT UNDERTAKINGS

Public responsibility for the relationship between major sectors of the
economy was taken for granted. It is not so clear why, in a competitive system, there should be any public responsibility for the efficiency of individual agencies or operators. There are two main
reasons.

The first stems from the fact that all railway services, about 75 per
cent of internal air services, about 40 per cent of public road passenger services and 10 per cent of road freight haulage services are
provided by nationalized concerns. Although the public corporations
were set up with the intention that they should be autonomous in

matters of day to day administration it was quite clearly expressed in the relevant statutes that they should be accountable to the public through the medium of Parliament. The duties laid upon the Boards differ in some respects from those of a normal commercial undertaking yet they are expected to fulfil these duties and, at the same time, remain financially viable. But duties which were mutually compatible in 1947 may cease to be so in the changed conditions of 1957 or 1967. It is therefore desirable that the terms of reference of the nationalized undertaking be kept constantly under review, though this should not be interpreted as advocating frequent changes which only serve to undermine the morale of the management.

Secondly, transport is a sector in which less direct Government intervention already has a long history. Almost every form of transport is subject to considerable statutory regulation. In road transport, for instance, the controls have been inherited from the early 1930s. The fact that the transport world has accustomed itself to this form of control does not necessarily prove either that it was originally justified or that it remains an economically desirable system of regulation. Again it would appear to be both sensible and correct to reconsider the controls from time to time to see whether they are still desirable in the interests of the efficiency of transport as a whole.

This, in itself, involves two major conceptual problems. In the first place we need to define efficiency for the individual operator and secondly we have to decide how traffic should be allocated between individual operators in such a way to maximize the efficiency of transport as a whole. The solution of this second problem we refer to as the co-ordination of transport.

Efficiency defined

What then, do we mean by efficiency in transport operation? The mechanical engineer, who gives the concept of efficiency its most precise meaning, defines it as the ratio of useful output to total input.

Both inputs and outputs are heterogeneous and can only be expressed on a single scale in money terms. But, in a purely accounting sense the total receipts are distributed between the factors of production with no residue, profit in this context simply being the valuation of the entrepreneurial function. Hence we would always find that efficiency was 100 per cent, which is clearly a tautology of no operational significance.

The output for the engineer cannot possibly be in excess of the input. For the economist exactly the opposite holds; the whole point of any economic enterprise is to add value to the material operated on in excess of the costs of the operation. This excess is the profit of the

operation. Might we not then redefine efficiency in economic operations as the rate to which this value is added, i.e. as a rate of return?
The rate of return would not be a figure on a finite scale like the
engineer's efficiency but would nevertheless be useful in comparing
operations.

Unfortunately the concept of a rate of return is extremely complex.
We need to decide the base for the rate, i.e. is it to be on turnover
or capital, and, if on capital, how should the capital itself be valued?
The potential for profits will depend in part on the degree of monopoly and any public restrictions on the commercial freedom of the
operator. For the multi-product operator the allocation of capital
costs may be indeterminate, and so on. Though we cannot at this
stage reject profitability as a measure of efficiency, we can see that
considerable subtlety of interpretation may be required. Thus the
need for a more simple method of assessing efficiency.

A host of simpler statistics of productivity have been produced in
both physical and value terms. This concept expresses output as a
rate, per unit, of one factor input. The Annual Accounts of the
British Transport Commission (B.T.C.) have included several statistical series of this kind such as tons per train, trains per mile of line,
receipts per worker. Some of these were submitted by the B.T.C. to
the Select Committee on Nationalized Industries in 1958. Similar
figures are available for other countries in the publications of the
International Union of Railways, the U.N., and the Economic Commission for Europe. At first sight it may seem that comparison of
these figures would give some useful comparison of the efficiency of
the railways involved. This is rarely so, however, for though the
statistics do provide a picture of the separate circumstances of each
railway system they do not denote relative efficiency.[1] Geography has
an important influence on the structure of the railway system and
hence on the statistics. The size and nature of a country, its natural
resources and the position and nature of the native industries all
affect traffic and the methods of dealing with the traffic. Moreover,
the fact that both inputs and outputs are heterogeneous renders the
statistics only moderately useful as measures of efficiency. A high
output per worker may be achieved by the utilization of a great deal
of capital equipment. A poor productivity in respect of ton-miles per
worker may be accompanied by a high figure for passenger miles per

[1] A useful selection of international comparative statistics is presented and
discussed by D. L. Munby in the *Bulletin of Oxford Institute of Statistics*,
February 1962, p. 113. The case against international comparison is cogently
argued by the B.T.C. in Appendix 47 to the Report of the Select Committee on
Nationalized Industries, July 1960.

worker. Even if both figures are low and there is no evidence of a great deal of capital equipment being used the efficiency of the line may truly be represented in terms of a high standard of service which can only be indicated by the consumer's willingness to pay. Thus the interpretation of these statistics is an art requiring intimate knowledge of the railway systems involved; they do not reveal by simple inspection any obvious and justifiable conclusions concerning efficiency.

We have now tried measuring both inputs and outputs in money terms, and measuring input in physical terms and output in money terms; neither gave a useful answer. To measure both inputs and outputs in physical terms would only be useful if both were homogeneous—but we have already demonstrated that inputs are not. Hence our only remaining hope seems to be to find a homogeneous measure of output the cost of which we can measure in money terms. If we succeed in doing this we can say that the most efficient situation or operator is that which produces our unit of output at the minimum cost. It is along these lines that rail enthusiasts have argued that if we take the ton-mile or the passenger mile as our unit of output it can be produced most cheaply by concentration of traffic on the railways in all except a small minority of cases.

This argument appears a little less telling when one considers the definition of the unit of output more closely. The distance of carriage and the nature of the commodity to be carried (its weight in the case of freight or its category in the case of passengers) are obviously important elements in the description of output but they are not the only ones. The customer may wish to be assured of delivery at a given time so that the service may be further described in terms of its speed and reliability; he may also wish to avoid damage or loss in transit so that the service may be further defined in terms of its safety. Thus the service produced has many aspects and the importance attached to each of these aspects will vary from customer to customer, from consignment to consignment and from time to time. In situations where it is possible to define precisely the service required it is in principle possible to compare the costs of various forms of transport in this respect (ignoring for the present the problem of joint costs) and to declare one form to be best suited or most efficient in transporting the consignment concerned. Far more common however is the situation where the customer is faced with several slightly different services, at different prices, between which he has to choose. In such a situation the most we can say is that from the consumer's point of view the most efficient service/price combination will be the one that he chooses. Thus it seems that there is really no problem, the market mechanism automatically selecting and supporting the most efficient operators.

This attractively simple conclusion is subject to two important qualifications. Firstly, though we have restated the operations of a market mechanism in such a way that price is only one of a number of characteristics of a proffered service, the relationship between cost and price still remains crucial. If the price obtainable for a service is below its cost of production this indicates that the community prefers the alternatives which could be produced with the resources concerned. Though difficulties arise when we try to apply this standard to individual traffics (the problem of overhead costs, discussed in the next two chapters), at least we can assert that for the individual producer securing custom does not in itself denote efficiency if losses are sustained in the process. Though this may seem to be a trite observation its implications are important. It means that, even for a nationalized concern which can carry on indefinitely despite incurring losses, a deficit is a sign of inefficient allocation of resources. The community is not getting what it values most from the resources used. Subsidies to private concerns, either open or hidden, have a similar effect.[1] Thus the first qualification is that the market mechanism can only guarantee an efficient use of resources, even in a perfectly competitive market, if the prices of services produced represent the total real cost to the community of the resources used in their production.

Secondly, the market mechanism has no way of indicating that a monopolist operator is not making the best of his productive potential. Nor will it automatically accept and support new forms of transport if they require a fairly large minimum scale for effective operation.

Thus the market mechanism can only select the most efficient from a large number of essentially similar operators in an unrestricted market where no subsidies are received. But the most difficult part of the transport problem is not to compare basically similar operators but to determine the roles to be assigned to various forms of transport with entirely different economic characteristics. This is the problem of co-ordination.

[1] We have assumed here that social benefit is directly related to effective demand. This is a value judgment. Subsidies might be justified by either of the following contradictions of this value judgment:
  (a) When it comes to spending his money the individual is a poor judge of what is really good for him. He thus needs some extra inducement to make the 'correct' choice.
  (b) The existing distribution of wealth is not conducive to maximum social benefit. A subsidy is justified if, with a 'better' distribution, this service would be in greater demand.

CO-ORDINATION

In the course of over a century of controversy in the transport field the word 'co-ordination' has frequently been used in direct opposition to the word 'competition'. This is strangely in contrast with the tradition of *laisser faire* economics which would regard them as practically synonymous.[1] Let us try to resolve this conflict.

The *Shorter Oxford Dictionary* defines co-ordination as 'putting into proper relation'. In an economic context at least two different interpretations of this definition are possible. Firstly, we may take it to refer to the establishment of a proper relationship between wants and their modes of fulfillment. On these lines it has recently been argued that, because freight transport is an ancillary service to industry, it is properly co-ordinated only when it is completely integrated with the rest of the productive process. Therefore, it is argued, transport on own account by traders is the best co-ordinator wherever it can be arranged. But the tailoring of transport to immediate and individual demands in this way may often prove to be inordinately expensive. In that case the provision of a large range of ready-made services is the best substitute. Free competition in public transport would therefore, on this interpretation, seem to offer the best prospects of proper co-ordination between transport and manufacturing industry.

The following alternative interpretation is more common. The various forms of transport may themselves be regarded as the entities which need to be set in proper relation to each other. Again, at first sight, competition would seem to fit the bill by indicating, through the consumer's choice, an order of preference and precedence between services. But, as we have already pointed out, this only holds so long as cost structures are simple and similar for all operators. If they are not, as in the comparison between road and rail transport, an unregulated market mechanism cannot be relied upon to ensure that the services required are produced at minimum real cost to the community. There are two ways in which we may attempt to overcome this difficulty. We may abandon the price mechanism entirely, replacing it by administrative direction of traffic in an attempt to minimize real cost. It is co-ordination by this means that has in the past been regarded as the antithesis of competition. But this is not the only possibility. We could, alternatively, investigate the differences between agencies and artificially compensate for them.

[1] See A. Plant, 'Competition and Co-ordination in Transport': *Journal of the Institute of Transport*, Vol. 13. No. 3, January 1932.

Thus, we could usefully retain competition between agencies as the co-ordinating mechanism.

In this book we shall be mostly concerned with this latter method. Even the Transport Act of 1947, which nationalized most of the public transport sector, required that the customer be left free to choose whether to send his goods by road or rail. An attempt was thus made to secure economies of scale which might not have been obtained under a simple market mechanism whilst at the same time retaining some freedom for the consumer to choose the service which suited him best.

The problem does not end here though. Not only has the chosen situation to be currently conducive to efficient operation but it must also promote the future efficiency of the transport system. Thus, whatever the form of control, we need to ensure that it is flexible. The best allocation of traffic between operators at present will not necessarily be the best in the future when the demands of consumers and the relative efficiencies of operators have changed. Moreover, neither existing firms nor existing modes are certain to remain always the most efficient so that freedom of entry is desirable in both respects. Great care must certainly be taken lest public control should ossify the existing structure.

Co-ordination, then, is not specific to a directed form of transport system but might equally well be achieved through a modified market mechanism.

PROTECTION OF SECTIONAL INTERESTS

So far we have considered public policy as entirely concerned with economic efficiency. There arise, however, from time to time, claims for the special treatment of certain minority interests irrespective of economic considerations.

(a) *Consumers*
There are many outlying areas for which the cost of maintaining a transport system of any kind exceeds the total revenue obtainable from it, either at present prices or at any conceivable price level. Traditionally both rail and road transport have been organized, as we shall see later, in such a way that services in these areas have been maintained despite their unprofitability. This has only been possible as the result of manipulation of the market mechanism so that monopoly profits might be taken on some services to pay for the subsidy to others. Many decisions have been made, both by individuals and by firms, on the assumption that the incentives offered by this

structure of transport prices would be maintained. Where irreversible decisions, involving long term commitments, have been made under the influence of this type of publicly controlled inducement, it has been argued that there should be a corresponding public responsibility to protect legitimate expectations against sudden changes in public policy.

In this respect, the Transport Users Consultative Committees have, under the provisions of the 1947 Act, sought to protect transport users where railway closures were proposed. Though they have not opposed all closures on principle, they have tried to ensure that reasonable alternative facilities are available when rail services are withdrawn. The B.T.C. accepted this responsibility and, where necessary, supported a replacement bus service by subsidy.[1] The 1962 Transport Act shifts the responsibility for supervising closures more clearly on to the shoulders of the Minister who will presumably continue to consider the traditional expectations of transport users most carefully in making his decisions.

(b) Employees

The Railways Act, 1921, set up a machinery to consider all matters of pay, hours of work and conditions of service of railways employees. In the name of fair competition this protection of employees was extended to the road transport industry by the Road Traffic Act, 1930, and the Road and Rail Act, 1933, though by virtue of the small scale of operation in road transport it has proved much more difficult to administer in this field.

The nature of this protection is at present undergoing considerable change. The reorganization of the railways has made it no longer possible to guarantee the continuation of employment to any individual. In view of this steps have been taken to provide compensation for those whose employment is terminated abruptly as the result of reorganization. Particularly in the cases of the railway workshops and of rural closures termination of employment may be a very serious blow if alternative employment is not available locally. The negotiated terms of compensation appear to indicate a progressive approach to this problem. It thus appears to be fairly well established that where a change in public policy causes hardship to employees they may expect some assistance in coping with their changed circumstances.

---

[1] By the end of 1962 thirty-four subsidies, amounting to £86,300 per annum, were being paid to bus companies.

(c) *The owners of transport concerns*

When menaced by increasingly intensive competition for traffic in the period of depressed trade in the early 'thirties the interests of both the railways and the larger road hauliers received some protection in the form of the new licensing system which enforced certain restrictions on entry into the industry.

Similarly, when nationalization was undertaken, it became a principle of compensation that the legitimate financial expectations of existing proprietors should not be arbitrarily shattered. Compensation was thus paid at levels sufficient to maintain pre-nationalization incomes for the proprietors.[1]

(d) *Manufacturers*

Motor manufacturers clearly have a vested interest in a solution of the transport problem which concentrates on road transport. As important exporters they tend to link their success in foreign markets, and hence the state of the balance of payments situation, very closely with the maintenance of a buoyant home demand for their products. This, they say, enables them to produce on the necessary scale to be internationally competitive; therefore the stimulation of home demand for cars is not only in their own but also in the national economic interest.

Subsidized transport facilities might also be consciously used as a means of giving a disguised subsidy to manufacturing industry, either to affect its location or to enhance its international competitiveness, in conditions where it is impossible or impolitic to offer a direct subsidy.

CONFLICT OF AIMS

We have so far suggested that public policy should endeavour to secure the economic distribution and use of resources in the short run and in the long run, both between sectors of the economy and within the transport sector itself, whilst at the same time offering some protection to the legitimate expectations of several minority groups with a particularly crucial interest in the outcome. This is an almost impossible task. Ensuring an economic allocation of resources will involve change which is bound to cause discontent for some injured interest. In the case of a railway closure, for instance, consumers will have to accept a different service, employees may be made redundant and in more serious cases there may be other multiplier

---

[1] See G. Walker and R. H. B. Condie, 'Compensation in Nationalized Industries' in *Problems of Nationalized Industry*, edited by W. A. Robson.

effects in the local community affected. In some way the advantages have to be weighed against the disadvantages, not simply to the operator but to the community as a whole.

Only if we redistribute the benefits of the change in such a way that everyone feels better off have we any comprehensive justification of the desirability of the charge. Otherwise we are faced with an intractable problem of welfare economics. It would be most unusual, however, if everyone were satisfied with a change in practice, so that this objective justification is unlikely to be achieved; indeed if we waited for conditions which gave complete satisfaction to everyone before change occurred we might still be living in caves! Therefore at some point an evaluation must be made of the relative gains and losses involved in a change. Value judgments are inescapable; there are some questions which, though appearing to be economic questions, cannot be answered solely in economic terms.

The problem may also be beyond the scope of static economic principles in a quite different sense. The aggregation of a large number of individual decisions may produce a result which none of the individuals concerned either foresaw or desired. For instance, it may be the present aim of most adults, or at least of most households, to run their own car. The complete fulfilment of these desires would necessitate a huge road building programme which might destroy some existing amenities. Such a process would be cumulative; as private transport expands public transport facilities lose custom and are therefore reduced, thus further stimulating the desire for private transport. It may also be irreversible. Areas cleared for the construction of motorways cannot easily be returned to their former state. The individual does not consider the contribution which his own decision makes to the ultimate outcome; he only considers its effects on him personally under present or immediately foreseeable conditions. Therefore, if the future of the transport system is left entirely to a market mechanism which does not take into account the full social implications of the individual decision an outcome may ensue which would not wittingly have been chosen. It is therefore desirable that the authorities accept the responsibility to investigate the long term implications of various types of transport policy so that we can at least plan to avoid the less desirable consequences.

CHAPTER TWO

Costs

Transport is not primarily an object of final consumption. Although there may be some who are prepared to spend money on rail or bus fares just for the pleasure of the ride, and many more for whom private motoring may still afford satisfaction as an end in itself, transport is generally an intermediate product. Thought of in this way, its function is to eliminate the temporal and spatial gaps between producers and consumers. The commodity in transit may be either a physical product moving from producer to consumer or a person travelling to work so that his labour (the product) may be available at the point required by his employer (the consumer of his labour). Or, if the product is not transportable, it may be the consumer who has to travel to the point of consumption.

Transport can be regarded as consisting of three elements. First of all a track must be provided over which movement can take place; 'track costs' may be taken to include signalling and policing facilities as well as the provision of the running surface.[1] Secondly, there must be a means of locomotion and carriage, usually provided jointly in one unit by air, sea and road transport but separately by rail. Thirdly, arrangements have to be made for the beginning and completion of the journeys; for collection and distribution, loading and unloading of the traffic, whether it be freight or passenger. These are referred to as terminal facilities.

Both the particular nature and the relative significance of these elements in the provision of transport vary from agency to agency with the result that services of differing character are produced. We may illustrate this point by comparing road and rail transport with respect to the following features:

  (i) Non-stop speed,
 (ii) Accessibility to users,
(iii) Frequency of service,
 (iv) Flexibility,
  (v) Bulk carriage.

---

[1] For air and sea transport where a natural highway is used this type of cost is a very small proportion of the total.

The greater potential for non-stop speed enjoyed by the railway stems primarily from the different characteristics of the tracks rather than from differences in the vehicles used. Although in the earliest days of railway development the railway companies provided the track and left operation to anyone who cared to hire it, the impracticability of such methods soon became apparent. The nature of railway track is such that facilities for overtaking are necessarily limited; its use therefore requires complex signalling and timetabling which can only be obtained by unified operation. Uninterrupted through running is thus possible by rail transport; this is not possible on the roads where speeds are limited by the fact that use of the roads is unrestricted and thus less susceptible to organized utilization.

On the other hand there are some 195,000 miles of road in the U.K. but only about 17,500 miles of railway route. Moreover, the process of transferring passengers or freight into railway wagons demands substantial specialist terminal facilities which can therefore be sited at only a limited number of points. Road transport terminal facilities, on the other hand, may be as simple as a bus stop or the parking space outside the house, shop or factory. Thus roads offer a far more accessible form of transport than railways due to differences in the nature and extent of track and terminal facilities.

The greater frequency of service and flexibility normally offered by road transport, and the advantage of rail with respect to bulk carriage, may be attributed primarily to the difference in size of the typical operating unit. The railway power unit is the locomotive, capable of hauling large numbers of detachable wagons or units of carriage, in road transport the typical unit combines locomotion and carriage and has a much smaller payload than the train. Therefore, for heavy flows, particularly of homogeneous, bulky traffic, the railway as a mass producer is likely to be at an advantage. For smaller consignments and more variegated traffic road transport enables a similar volume to be distributed more evenly through time and permits the provision of services more specific to individual requirements.

Thus, although track, locomotion and carriage, and terminal facilities must, be provided in both road and rail transport, technical differences result in the production of services with different characteristics. It must be emphasized, however, that the distinctions so far drawn are of the most general nature and that the significance of the relative advantages that have been ascribed to road and rail transport will vary greatly from one transport user to another. In the light of their own particular requirements users will weigh the relative advantages of road and rail. Their distribution of traffic between the alternative forms of transport will tend to be that which minimizes the total costs of production and distribution. This may be quite different from simply minimizing the money outlay on transport. A saving in breakage losses resulting from superior handling facilities or a reduction in inventory costs resulting from the absence of demurrage penalties may more than offset a marginal price differential. A transport user may therefore choose a particular form of transport even though it appears to be more expensive than the alternatives.

Having admitted that money cost is not the only element in user's choice does not detract from the fact that it is a very important one. For the rest of this chapter we shall be considering the nature and magnitude of various categories of cost for road and rail transport.[1]

RAILWAY COSTS AND COSTING

The Annual Report and Accounts of the Railways Board (formerly of the B.T.C.) provide a wealth of information concerning the expenses of running the railway system in the previous year. Information in this form is of limited use, for two reasons:

(i) The information is 'ex post', whereas ideally the operator wants it *before* he provides the service;
(ii) The information concerns aggregates, whereas ideally the operator wishes to know about the costs of providing *particular units* of service.

It is of course instructive to know whether a service or line has proved profitable in the past, but it is infinitely more useful, if profitability is the aim of the concern, to have costings information in advance so that unremunerative traffics may be avoided.

It is only since 1951, when the Traffic Costing Department was established, that the railways have acquired much knowledge of this kind.[2] The functions of this department were to study and present suitably analysed information concerning the operating costs of the various B.T.C. undertakings in order to increase the knowledge and understanding of costs amongst those responsible for the commercial and operating sides of the business.[3]

---

[1] Similar details for air, sea and pipeline transport are contained in Chapter 11.
[2] For a description of the service see Report of the Select Committee on Nationalized Industries, 1960, Appendix 39.
[3] The increased significance attached to traffic costing is indicated by the fact that the initial small headquarters and regional staffs have since been supplemented by the appointment of Traffic Costings Officers at Line, Division and District levels, in each region.

Traffic Costings Service (T.C.S.) analyses have often been made public so that we probably have as much information about the costs and costing techniques of the railways as of any other enterprise in the country. Firstly, the Annual Reports of the Commission have from time to time referred to costing studies made during the year. And secondly, over the period since nationalization the obligation imposed on the Commission to submit charges schemes to the Transport Tribunal, however undesirable on other grounds, has proved a fertile source of information for outside observers. Detailed cost analyses have been presented to the Tribunal to support the Commission's applications and these have been explained and supplemented by an understandably reluctant Commission under cross-examination.

For particular years, e.g. 1955–6, we thus have a detailed breakdown of the costs of railway freight transport for traffic carried in 'adverse but not extremely adverse' conditions[1] and also for traffic carried in 'normal' circumstances. In addition, much information was presented about the ranges of cost which the railways encountered in their operations and about the costing techniques on which the Commission's calculations were based.

We shall not here be concerned with a detailed analysis of this information,[2] but rather with examining the railway operator's general approach to the principles of costing in the light of the practical complications of railway operations. Before doing this, however, it may be useful first to state the economist's approach to the problems of costs and costing.

The economist's approach to costs

There are two basic notions in the economist's approach to costs. Firstly, real costs for the economist are 'opportunity costs', i.e. the cost of any productive process is the most desirable alternative that could have been produced by the resources used. This is expressed in money terms through the market valuation of the factors employed. Nevertheless it may differ from the money expenses incurred as it does not include any tax element or any monopoly profit or economic rent secured by the suppliers of factors of production. It will not depend on the treatment of depreciation and managerial salaries. Secondly, there is the concept of escapability. The opportunity cost of a particular unit of output, however large or small, is to be found

---

[1] Defined in such a way that only 10 per cent of railway traffic was carried at higher levels of cost.

[2] For more details see C. D. Foster, 'Some Notes on Railway Costs and Costing', *Bulletin of the Oxford Institute of Statistics*, Vol. 24, No. 1, February 1962.

C

by asking what costs may be escaped if we do not produce this particular unit.[1]

Now let us apply these concepts to railways. Suppose, to begin with, we ask what saving in resources can be effected by carrying one less passenger from X to Y. If exactly the same train service will be provided whether this particular passenger travels or not, the answer to our question will be 'very little indeed'. In fact the saving is likely to amount to little more than the cost of the cardboard in a ticket. Now this may be an interesting item of information about the operator's cost structure. But it must be emphasized that no particular conclusion follows concerning how much that passenger should be charged. Nor should we be misled into thinking that near-zero immediately escapable cost is a peculiar feature of the transport industry. The costs escapable by printing one less copy of a newspaper or by making one less packet of detergent or indeed of making one less unit of anything where the raw materials form a small proportion of total cost are similarly small.

Underlying this is the fact that many inputs are fixed in quantity, either for particular time periods or over certain ranges of output. Labour costs, for example, are generally temporarily fixed. Although the quantities *needed* may be finely variable with output, in practice labour has to be hired for minimal periods of a day, a week or a month. The other sort of fixity in costs is technical. To carry a single passenger will require the laying of railway track which may then have a technical capacity for the carriage of a very large number of further passengers. There is an irreducible minimum input of many factors which, once employed, may be sufficient to cope with a larger volume of output without further variation, i.e. there are significant indivisibilities.

To ask what resources may be saved by discontinuing a service is therefore ambiguous. Our original answer was confined to those savings which would accrue *immediately*, and from not providing the most marginal unit of output, an individual passenger journey. Very different saving of resources would be achieved if we widen the application of the escapability criterion to enquire:

(i) What savings would be effected if not only a single additional passenger was not carried, but if, say, the 10.00 a.m. service from X to Y was discontinued altogether? Or if *all* passenger services between X and Y were discontinued? Or if the X—Y line were completely closed down?

---

[1] See W. A. Lewis, *Overhead Costs*, Chapter 1.

(ii) What costs would be avoided in each of these cases within a
single day? Or a week? Or a year?

Escapability of costs thus has both a temporal and a technical
dimension. There is not a single simple division between escapable
and inescapable costs but rather a spectrum of escapability varying
with the time period and the range of output which we consider
appropriate.

Mr C. D. Foster has attempted a division of railway freight transport costs on the basis of their temporal escapability. He shows that
over half of the railways' costs must be met in the short run if the
railways (or some part of them) are to be kept running at all. About
10 per cent could, roughly speaking, be avoided for a period of perhaps one year before the railways were no longer safe to operate,
whilst for about a quarter of the total the renewal of assets could be
escaped for a longer period before the system would finally become
unworkable.[1]

Similarly, with regard to technical fixity, the 'lumpiness' of factor
inputs leads to varying degrees of escapability according to how the
service under consideration is defined. This may result from the
existence of either joint costs or common costs.

Joint costs occur where the provision of a certain service necessarily entails the output of some other service. For example, if a
train based on X and running to Y has to return directly to X, the
return journey Y—X is jointly produced with the outward journey
X—Y. If it is then asked which costs are avoided by discontinuing
the Y—X service, the answer may once again be near-zero. Joint
costs are only jointly escapable—if neither of the services is provided.

Common costs may be distinguished from joint costs in that,
although they must be incurred if a certain range of output is to be
entered at all, no individual unit of output within this range necessarily entails the production of any other unit. For example, the
overhead costs of a particular stretch of track (X—Y) may be common in three senses.

(a) To all units, whatever their nature, using the track.
(b) Both to traffic originating in X or Y and terminating in the other,
and to traffic which uses the X—Y line only as a section of
longer journeys.
(c) To both passenger and freight services (except in so far as they
require different standards of track).

Escapable cost is thus only meaningful when stated in a particular
and precise context.

[1] C. D. Foster, *op. cit.*, p. 89.

Finally, an extreme case will illustrate how the economist's approach differs from the traditional accounting approach to cost. Consider a newly constructed and expensive asset which is both highly specific and infinitely durable (a railway tunnel for instance). As the asset is specific it has no transfer value in other uses; as it is infinitely durable its use implies no additional call on resources at any time. Nothing is escapable by a decision not to use it once it is in existence so from the economist's point of view its use is costless.

Direct and indirect costs

This is the basic distinction which railway operators themselves make in their analysis of costs. Even in railway circles the terms have not been used in a unique way; nor, in their most common railway usage, do they correspond simply with the concepts we have discussed hitherto. The distinction between direct and indirect costs has, in fact, been made in two entirely different ways:

(a) *The accountant's distinction.* In the B.T.C. financial accounts the working costs of the railways are regarded as direct costs whilst the indirect costs are simply the proportion of central charges allocated to the railways. These latter consist mainly of central administration and interest charges. This accounting definition is of no great significance either to the economist or to the railways Traffic Costing Service (T.C.S.) which has its own peculiar use of the terms. Unfortunately, it is difficult to ascribe any clear economic meaning even to the T.C.S. distinction.

(b) *The Traffic Costings Service distinction.* In giving evidence to the Transport Tribunal in 1955, the head of the Costing Service, Mr A. W. Tait, described the distinction thus—'direct costs cover all those items, the cost of which, taking a sufficiently large volume of traffic and a sufficiently long term view, may be expected to vary with changes in traffic: indirect costs cover items which, in general, do not so vary'. This appears to be a distinction between fixed and variable factors which accords fairly well with the economists' outlook. Of course, in order to give it a precise meaning in terms of escapability, it would need to be placed in a particular context with the volume of traffic and the period of time defined. On this definition of direct and indirect costs it does not make sense to qualify the categories except in the context of particular assumptions.

However, handled with caution, this distinction does appear to be one which could be used as a guide for commercial policy.

One of the problems of defining escapable cost was the result of indivisibilities. So in an attempt to make the escapability definition more precise the T.C.S. tried to incorporate within it some treatment of common and joint costs. They suggested that direct costs could be imputed to particular transactions while indirect costs could not. As an example Mr Tait pointed out that in respect of 'C' lines (track predominantly used by goods traffic, but also by some passenger trains) track costs were all allocated between goods traffic as they would not be reduced if the passenger traffic had not used the track. But this argument could have been pursued further to absolve any particular freight consignment from sharing the track costs on exactly the same reasoning.

In giving evidence to the Select Committee on Nationalized Industries in 1958 the B.T.C. committed themselves even more thoroughly to a distinction in this vein: ' . . . Direct costs of operation based on fairly firm foundations of fact and so-called "indirect" costs which can only be apportioned arbitrarily.'[1] Indeed, their concern with the principles governing the apportionment of apparently indivisible costs seems to lead them into conflict with the escapability criterion. They even argue explicitly: 'The direct costs ascertained by traffic costing methods are not the same thing as short term marginal costs. Nor do they correspond with the savings that would flow immediately from the discontinuance of a small part of railway activities. To arrive at assessments of this kind is a separate process.'

In practice, the railways have identified certain general headings of cost as comprising the indirect and direct categories. The costs of general administration and all track and signalling (with the exception of that attributable to terminal stations and marshalling yards) are regarded as indirect and the residue as direct. On this basis, some 44 per cent of total costs are indirect and 56 per cent direct. As we have seen, the railways do not believe that all the direct element can be imputed unequivocally to 'small parts of railway activities'. Indeed, as Mr Foster shows, an interpretation in this vein, using the individual wagon load as the basis of calculation, would result in only 16 per cent of total costs being classed as direct. The remainder would, in one way or another, be 'apt for probabilistic accounting'.[2]

It seems, therefore, that what the railways have done is to draw a line between different degrees of arbitrary apportionment. The

[1] Report of the Select Committee on Nationalized Industry. Appendix 39, p. 450.
[2] C. D. Foster, op. cit., p. 89.

'reasonable foundation of fact' may be that at least costs are being statistically allocated over similar units of the same kind of output, whereas to try to impute administration and track costs would imply a quite different order of arbitrariness in that allocation would have to be made between passenger miles and ton-miles, i.e. altogether different products.

In pursuing this line the T.C.S. have confused rather than clarified the issue. Direct cost ceases to be interpretable either as the costs escapable for a small change of output (because of the arbitrary allocation of certain common costs) or as those escapable for a large change (because it does not include certain track and signalling costs which may be escapable in such a context). Consequently it does not give any guide as to immediate commercial policy or to longer term planning of the facilities to be provided.

In practice the railways seem to adopt a much more pragmatic and flexible approach than their own theorizing would suggest. For instance, even in their treatment of track costs they do attempt to assess the costs of the different categories of track and the allocation made between passenger and freight traffic does seem to be based partly on the escapability criterion. The T.C.S., moreover, shows a healthy dislike of averaging, particularly where the multiplicity of products would require the arbitrary comparison of entirely different output units. They showed considerable reluctance to submit the track cost figure for passenger and freight running to the Select Committee, probably because they were aware of the substantial variations between costs in different situations. In Appendix 39 to the Report of the Select Committee it is made quite clear that the T.C.S. attempts, when costing a traffic, to take figures for the items involved which relate to the specific time, place, type of traffic and conditions of service involved in transporting that particular load. Moreover, the operations are broken down in considerable detail.

Now from the economist's point of view the 'real' cost of any particular operation cannot become apparent simply by analysis of the operation itself unless we also know the conditions of availability of all the factors concerned. Only in this way can we tell whether the use of a crane, a platform, a section of track or a labour hour, represents any lost opportunity or not. Henceforth in this book we shall use the term direct cost to refer to those costs directly escapable in the context of a particular decision and not simply to refer to those costs categorized as direct by the T.C.S. The essence of indirect costs is therefore that there is no objectively unique way of allocating them. The use of averages as estimates for direct costs which cannot be traced in every individual instance may be a very necessary element in costings procedure, but only as an approximation to the true direct costs. In practice railway management shows considerable awareness of this despite the failure to express it satisfactorily to either the Select Committee of the Transport Tribunal.

THE COSTS OF ROAD TRANSPORT

Many of the complications which arise in railway costing are not so important in the road transport sector. From the operator's view-point costing is simpler in road transport in three main respects.

(1) The typical operating unit, the lorry or bus, is smaller than the railway train. Furthermore this unit combines locomotion and carriage. Hence the problem of costs common to a large range of output is substantially diminished. In road haulage, for instance, the convenient costing unit, the lorry, will much more frequently correspond closely to the charging unit, the load, than in rail transport.

(2) The typical administration unit is also smaller than in rail transport with the result that records can be kept more manage-ably. Moreover, these records are fairly simple, consisting of drivers' log sheets and fuel and repair bills.

(3) From the operator's point of view the allocation of track costs presents a less serious problem than in rail transport. Use of the roads may effectively be regarded either as free or as paid for directly in the taxation on fuel and licence fees. But these pay-ments may be specifically costed to individual vehicles and, in the case of fuel taxation, even more finely to specific traffics in that they are escapable if the load is not carried. Hence the element of joint overheads is reduced.

The costs of road haulage, for example, can be divided into a few main categories (see Table 1). The relative importance of these categories will vary from operator to operator. Our hypothetical example shows the proportions for a small operator with relatively low administration and depot overheads.[1]

Fuel costs obviously vary in direct proportion to the mileage run. They also depend, though with less sensitivity, on loading and on the nature of the route traversed. Drivers' wages and expenses are escap-able in the short run if they are employed on a short term basis or if alternative traffic is available. Similarly, as depreciation of commer-cial road vehicles is fairly directly related to use and not simply to

[1] For the cost structure of a larger road haulage concern see Table 12, page 139. For that of a large passenger transport concern see Table 5, page 110.

TABLE 1

Road Haulage Cost Structure

Variable Costs		%
Drivers' wages and expenses	33	
Fuel	17	
Tyres and repairs	18	
Depreciation	10	
	—	78
Fixed Costs		
Licences and insurance	5	
Administration and depot costs	17	
		22
		—
		100

the passage of time, this also is escapable in the short run by leaving the vehicle idle.[1] In the case of tyres and repairs the costs vary fairly directly with the use made of the vehicle, for, though the actual expenses occur infrequently and the increment of tyre wear or vehicle wear is not discernible, it can be related sensibly on average to vehicle use. Indeed, it is fairly common practice for tyres to be supplied on a contract basis at an agreed rate per vehicle mile, thus making tyres a directly variable cost.

About three-quarters of road haulage costs vary directly with vehicle usage and hence can be attributed directly to the particular vehicle loads carried. This may still, of course, leave a problem of allocation in the case of the remaining quarter of total costs and in the case of less than full load consignments but such problems are small compared with those encountered in railway costing.

Utilization and load factors

We have seen that many of the complications in railway costing have no parallel in road transport. There is one costings problem, however, which affects all transport agencies to some extent.

// Costs vary not only with the volume of traffic, the type of traffic and the physical conditions in which it is transported; they also depend on when the service is provided. As a result of the immediate perishability of transport the cost per unit of service sold is inversely proportional to the degree to which capacity is utilized. Hence we can distinguish between the *potential* costs, which are the unit costs

[1] There is some disagreement amongst cost accountants as to how depreciation of road vehicles should be treated. Wheldon's *Cost Accounting and Costing Methods* treats it as a fixed cost whilst Biggs' *Cost Accounts* treats it as a variable. For an economist's view of the problems of depreciation see G. J. Ponsonby, 'Depreciation with Special Reference to Transport', *Economic Journal*, Vol. 66, March 1956.

which would prevail if all capacity could be fully utilized all the time, the *actual capacity* costs, which would be the costs relevant if the capacity actually offered were fully utilized and the *actual* costs per passenger or ton-mile which are the costs per unit for the actual loading achieved.[1]

It would be physically possible to utilize either a railway train or a road vehicle for up to 18 hours a day, 6 days a week, for up to 48 weeks in a year, the rest of the time being taken for cleaning, refuelling and maintenance. On these assumptions the railway train would be likely to have the advantage because of its greater capacity and higher average speed capability. But these are only the potential costs. This maximum utilization of capacity is unattainable in practice due to scheduling problems and the 'peakiness' of demand. In fact, rail capacity suffers more chronically from under-utilization than road transport capacity so that the rail advantage in terms of cost per seat mile offered or per capacity ton mile is considerably less than that in potential costs.

Furthermore, not all units of capacity provided can be transformed into units of service sold. The difficulty of obtaining custom equal to the payload of the transport unit used is known as the load factor problem. Once again the effect is more severe on rail transport than on road because the railway train is less flexible than the road transport unit when it comes to matching supply to fluctuations in demand. In long distance passenger transport this difference has been exaggerated by the licensing system which ensures high load factors for the road operator by restrictions on competition.

Much, if not all, of the railway potential cost advantage is thus lost in practice due to the difficulties involved in securing satisfactory utilization of capacity, which affects the railways much more seriously than road operators.

PRIVATE COST AND SOCIAL COST

So far we have been talking exclusively of the expenses incurred by operators in providing a particular facility. These are not necessarily all the costs which society incurs as a result of the operation, however. The railway train may produce smoke or fumes with deleterious effects on the health of those living in close proximity to the railway. The heavy lorry unloading in a main street may cause a traffic jam resulting in a waste of time and fuel for a large number of

[1] See D. M. Dear, 'Some Thoughts on the Comparative Costs of Road and Rail Transport', *Bulletin of the Oxford Institute of Statistics*, Vol. 24, No. 1, February 1962.

other road users. In both of these cases the costs fall not on the operator himself but on some other unfortunate section of the community. The costs actually incurred by society as a whole as a result of any activity, whether incurred as private expenses or not, are known as social costs.[1]

The difference between private costs and social costs may be negligible as, for instance, in the case of a vehicle running on an uncongested country road. But it is not always so. If it is not the individual operator may be encouraged by the price mechanism to undertake operations the real cost of which exceed the total benefits accruing to the community. Thus when we consider, in later chapters, the use of the price mechanism as a co-ordinating influence we must remember that the total cost incurred by the community as a result of a particular operation may be very different from the private expenses on the basis of which the entrepreneur would make his decisions.

[1] The relationship between private and social product from which the distinction between private and social cost is derived is expounded by A. C. Pigou: *The Economics of Welfare.*

CHAPTER THREE

Pricing

In this chapter we shall discuss the principles on which the pricing of transport should be based. It might be useful, however, to begin by showing why transport charges should be a concern of public policy at all. The Government does not lay down pricing rules for the producers of soap, or haircuts, or television sets. Why then should the price of transport be of such special interest?

Public utilities are often said to require special attention because of their monopoly position. It must be emphasized at the outset that State interest in the price policy of transport operators is no longer simply an attempt to prevent the exercise of monopoly power. Control may have originated for this reason but transport today is a sector in which a high degree of competition prevails not only between the various public transport agencies but also between public and private transport. Thus even in London where public ownership incorporates both rail and bus facilities private motoring provides a check on the abuse of monopoly power in public transport.

Indeed, it is just this competitiveness, arising from the substitutability of road for rail and private for public transport, which makes pricing policies so relevant to the problem of transport co-ordination. There are few traffics which travel by road or rail which could not alternatively be carried by the other. In such circumstances it is the fact that prices *count*, that they are an important consideration in determining the distribution of traffic between agencies, which makes it vital that they should be sensibly based. It is normally assumed that, in a competitive situation, the profit motive will be sufficient to cause operators to relate their price to their costs of production in such a way as to secure that consumers' requirements are satisfied with the least possible call on scarce resources. This is what we mean by saying that prices are sensibly based. Transport pricing is of special interest because there is reason to believe that there are complications in this sector which would prevent competition from bringing this about automatically. The variation between agencies of the proportionate significance of overheads, the immediate perishability of the product, the differing circumstances under which track is provided and the various social policies which have traditionally been forced on the transport sector combine to create special problems.

The question of the social costs of road use and the effects of social obligations on the division of traffic between agencies will be considered in later chapters. In this chapter we abstract from these complications and consider the implications of various pricing policies in a sector where agencies with widely differing private cost structures are in competition.

PRICES AND COSTS

We have already stressed that transport services are multi-dimensional in character. The lowest priced haul is therefore not necessarily the cheapest from the point of view of the transport user who will also take into account the additional costs implied by differences in speed, reliability, accessibility, flexibility and so on. Nevertheless it is still important to devise a structure of transport prices which allocates traffic in such a way that for the provision of a required service the real cost to the community is at the lowest possible level. There may be alternative methods of pursuing this aim, involving the administrative direction of traffic; we shall concentrate, however, on the possibility of achieving it by the price system.

Since transport users choose their agency at least partly on the basis of price it is important that charges should be related to the real value of the resources used in providing a service. Only if the value of the factors used is reflected in the price charged will individual users' choices be translated into the most economical employment of resources in the transport sector as a whole. Can we not therefore merely instruct transport agencies to charge according to cost?

Unfortunately we cannot. The directive to charge according to cost is highly ambiguous and can serve only as a statement of the problem rather than as a solution. The outstanding difficulties to which such an instruction gives no unique solution may be approached in two stages. Firstly, we must decide which of the operator's expenses and, possibly, what else in addition, should be covered by his charges. Secondly, we must decide how these costs should be recovered since there are various pricing techniques from which to choose.

Marginal cost pricing
We may begin by considering the general pricing rule for public enterprise in a mixed economy developed independently in the 'thirties by Professors Hotelling[1] and Lerner.[2]

[1] H. Hotelling, 'The General Welfare in relation to the problem of Taxation and of Railway Utility Rates', *Econometrica*, Vol. 6, No. 3, July 1938.
[2] A. P. Lerner, *The Economics of Control*.

The Hotelling–Lerner rule states that only marginal costs of production need to be taken into account and that recovery of such costs should be effected by simply instructing public enterprises to charge a price equal to marginal costs.

Implementation of the rule will frequently imply a failure to cover total costs; this can occur both in cases where price is made equal to short run marginal cost and in cases where it is equal to long run marginal cost. Firstly (see Figure 1) is the case where mar-

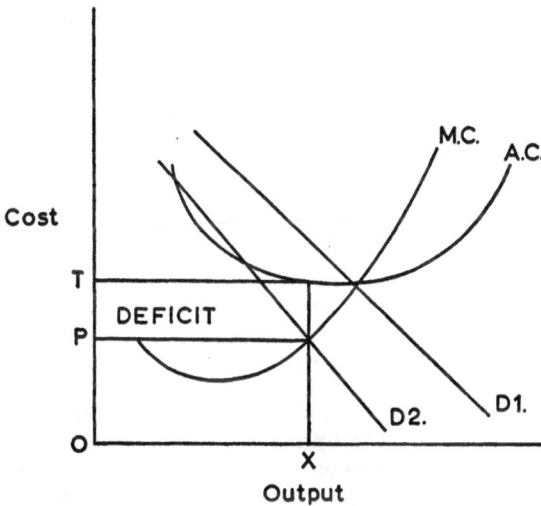

Figure 1

ginal costs are rising in the short run. If output is extended to the point at which marginal cost (MC) equals price and the demand curve cuts the short run average cost curve to the left of the point of minimum average costs a deficit must occur. In our diagram, demand D_1 would produce a break-even by the application of the rule. With demand curve D_2 price is equal to marginal cost at the level OP for an output of OX. But at this level of output the average cost per unit is OT and a deficit of PT.OX is incurred.

Secondly, where decreasing long run average costs are experienced, long run marginal cost must be less than average cost. In these circumstances even long run marginal cost pricing must result in a deficit (see Figure 2).

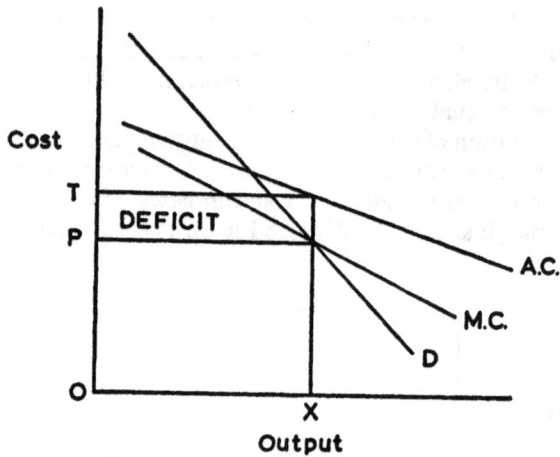

Figure 2

// There will thus be a financial shortfall both,

   (a) If short run marginal costs are charged where plants are running below capacity in the short run.[1]

   (b) If long run marginal costs are charged in concerns operating under conditions of long run decreasing costs.

The Hotelling-Lerner proposition is that we must not be hidebound by traditional accounting conventions in determining the policies of public enterprises and that financial losses should be made good from general taxation, including any surpluses of other public enterprises.

Now let us consider why the Hotelling-Lerner rule is advocated. The essence of the argument is that prices should reflect opportunity costs. To secure the best use of scarce resources the price with which the consumer should be confronted must indicate the value of alternative output which has to be foregone as a result of the decision to consume the product in question. Anyone prepared to meet the marginal costs incurred has indicated that he values that product of the additional resources thus employed more highly than the market values the alternatives and thus should not be denied the right to consume the product. This, it is argued, will lead to an optimum allocation of resources and is hence clearly a goal at which public policy should aim.

This analysis is open to a number of serious criticisms.[3]

[1] i.e. below minimum average cost for the existing capital equipment.
[3] See I. M. D. Little, *Critique of Welfare Economics*, Chapter II.

(1) The argument implies the application of marginal cost pricing throughout the economy. If it is not commonly applied there is no particular merit, from the point of view of overall allocation of resources, in applying it to specific sectors.[1]

(2) If financial losses, sustained through the application of the rule, were to be made good by general taxation, it would have implications for the distribution of income. A value judgment is thus implicit in the rule that consumers of the products of some industries should be subsidized by other sections of the community. Or, to put it more generally, 'that certain long run opportunity costs for the community as a whole should be ignored in the interests of the greater short run utilization by consumers of specific factors of some stated degree of durability'.[2]

(3) It is not clear whether short run or long run marginal cost should be charged. Even if it were clear, difficulties of interpretation could still arise where joint variable costs occurred in a sector with a non-homogeneous output.

(4) The rule does not give any guidance at all concerning investment criteria, whilst automatically destroying the guide of profitability commonly used.

(5) Not only does profit act as an indicator of the relative merits of current projects but also as a stimulant to new developments. This also might be destroyed by the application of the rule.[3]

In face of these criticisms it is tempting to conclude that the case against marginal cost pricing is overwhelming and hence dismiss it as totally irrelevant to the problem of transport co-ordination. It has been suggested, however, that marginal cost pricing might still be of use in determining the allocation of resources within an individual sector of the economy. For instance, E. J. Mishan argues[4] that if the marginal cost rule is used by all concerns within a specialized sector, say transport, then, although the share of national resources used in that sector may be too large or too small, at least the utilized resources are properly allocated within the sector, ignoring external effects.

It is thus argued that marginal cost pricing might be specifically

[1] See R. G. Lipsey and K. Lancaster, 'The General Theory of the Second Best', *Review of Economic Studies*, Vol. 26, No. 63, 1956.

[2] J. Wiseman, 'The Theory of Public Utility Price—An Empty Box', *Oxford Economic Papers*, Vol. 9, No. 1, February, 1957.

[3] Though it has been argued that such stimulus may be provided by administrative means. See G. C. Means, *Pricing Power and the Public Interest*, especially Chapter 18.

[4] E. J. Mishan, 'Welfare Criteria for External Effects', *American Economic Review*, Vol. 51, No. 4, September 1961.

applicable in sectors relatively isolated in effect from the rest of the economy. We must now consider whether or not transport is such a sector. We argued in the opening chapter that the elasticity of demand for transport as a whole is not likely to be great in the short run so that marginal cost pricing is not likely to give rise to very much immediate substitution of transport for other commodities or services.

But this is dealing only with the immediate problem of the use of existing facilities. It does *not* give any guidance as to what services should be provided. In fact, the crucial problem for the transport operator may often be to decide what services and capacity to provide rather than merely to devise a pricing system to make fullest use of existing capacity. Variations in the output of a transport enterprise may not be smoothly matched by the employment of factors of production because many of the inputs are of a 'lumpy' nature. To carry even a single unit of traffic will require a minimum of staff, fuel, rolling stock and track. Once this is provided many more units may be carried before a further truck need be added, substantially more before a further locomotive is required and a great deal more than this even before it becomes necessary to lay more track. Staff, rolling stock and track are all technically indivisible inputs to some degree or other. Marginal cost in such circumstances becomes similarly discontinuous. The last unit of traffic carried on a track reaching full capacity may incur nearly zero marginal cost. The addition to total costs incurred by carrying a further unit will be enormous, amounting to the whole cost of laying new track plus increments of other factors to work it. Implementation of the marginal rule in these circumstances is obviously problematical.

Professor Lewis argues that confusion on this point has been due to a tacit acceptance that the unit of product is the unit on which the calculation of economic cost must be based.[1] Marginal cost is thus often taken to imply a quite arbitrary limitation to the costs which are immediately escapable for the smallest unit of output. But there should be no greater significance attached to the passenger than there is to one of the other indivisible units required for his conveyance. The decision not to carry one additional passenger may yield negligible savings in resources. The same may be said of one more book, or any other product where indivisible inputs are important and are not used to capacity. But the decision not to run a particular train or service, or to close a line completely, all imply increasingly large savings in resources. All of the resources released, in so far as they can be put to other uses, involved some opportunity

[1] W. A. Lewis, *Overhead Costs*, Chapter 1.

cost in their railway use despite the fact that the savings could only be achieved over a longer time period or for decisions of a relatively large scale.

The requirement that price reflect opportunity cost thus does not simply imply that the price be made equal to short run marginal cost. On the other hand, we still cannot argue that opportunity cost is identical with accounting cost for four reasons:

(i) 'Economic' or 'opportunity' cost is dependent on the size of unit and type of decision in question.

(ii) Economic cost would normally be considered as replacement cost whereas accounting cost is usually calculated on a historical basis, giving a downward bias in inflationary periods. On the other hand if the supply of a factor is very inelastic that factor may be able to earn a quasi-rent. Thus, though for the individual operator the price paid for the factor (private cost) does involve lost opportunities, to the economy as a whole the opportunity cost is the value of production which would have been achieved by the next most profitable use of those resources. This may in fact be considerably less than the current market valuation of the factor. In this case expenses include an upward bias in comparison with opportunity costs.

(iii) To the extent that assets were installed to meet an expected demand which has not materialized there is no question of replacement. But, under normal circumstances, the reason for charging for the use of specific assets is to have a guide to replacement policy. In this case therefore such a charge ceases to fulfil any function. Neither at present nor in the future is any alternative foregone as a result of permitting use of the asset. Thus it has no economic cost.

(iv) A similar argument may be applied to non-replaceable assets such as railway tunnels. At the time when a permanent asset is under construction opportunity costs are involved as the factors employed could be used to produce some alternative product. But the moment construction is completed the asset ceases to have any economic cost since it is specific, non-transferable and non-renewable.

A decision not to charge for an asset falling into the last two categories, though permitting the extent of utilization of the asset to be increased without any effect on other resources, either at present or in the future, is not without implications on income distribution. The asset would not have been constructed in the first place unless it was anticipated that users would be prepared to pay sufficient in charges

D

to cover its construction costs. This implies that in future years they would be willing to forgo the consumption of other products to enjoy use of the asset, thus permitting a greater proportion of scarce resources to go to non-users of the asset. If the price they are subsequently charged does not cover the cost of construction then over the period taken as a whole a redistribution of income has taken place from non-users to users of the asset.

Short run marginal cost is not necessarily synonymous with economic or opportunity cost. But even if it were, marginal cost pricing could still not be considered ideal because it is sometimes impossible to separate matters of economy from matters of equity.

We shall now consider, on a more pragmatic basis, the effects of various types of pricing policy for transport, taking first the question of which costs are to be covered and subsequently the problem of how they are to be recouped.

Which costs should operators try to cover?
In contrast to the marginal rule, which we have already considered, the principle that 'he who benefits should pay' is frequently used as the basis of an argument for covering total costs. To the extent that a publicly-owned industry fails to cover its total costs, its financial losses must be made good by external subsidy. It is sometimes argued that railway losses may be met from general taxation without infringement of the 'he who benefits must pay' principle, since taxpayers are generally transport users as well. But this argument is only of limited validity. There are many taxpayers and transport users who never use *public* transport; moreover the system of general taxation is certainly not based on the degree of benefit conferred on individuals by the operation of a public transport system. Local taxation might be a little more precise in this respect, as costs specific to certain areas might be isolated and recovered through the rates. Nevertheless, the difficulty of identifying beneficiaries, and of devising taxation techniques to ensure that it is they who are covering losses, is likely to be insuperable, so that any failure to cover costs will infringe the principle of payment by beneficiaries.

A second argument for covering costs arises from the fact that the transport industry is a mixture of public and private enterprise. In the private sector full costs have to be covered if the operator is to remain solvent. Moreover, many operators also aim at a high degree of self-finance for development. Though self-finance is not necessarily desirable, nevertheless if the public sector is not pursuing a similar policy then transport consumers are unable to choose between alternatives on a like with like basis. Their choice in such circumstances may not lead to a particular service being provided at the minimum real cost since prices will be indicative of the different constraints under which operators are working rather than of the resources which they employ to produce their services.

Total cost coverage may (but not necessarily must) provide some check on the efficiency of the public corporation. Where there is no yardstick of profit to measure success there is a danger that the incentive to reduce costs will also disappear. Even where this problem has been largely solved by administrative methods, it is still possible that Parliamentary and public opinion will ignorantly equate inefficiency with the process of subsidization itself, despite the low price of the product. Constant carping by a public conditioned to the values and criteria of private enterprise may have such a disturbing effect on morale that eventually the concern does give cause for justified complaint. In such circumstances it may well be in the long term interests of both the particular concern and the economy as a whole that it be set an objective which is both attainable and understandable to employees and to the public.

Now, in contrast, let us look at some possible justifications of subsidies. One argument for covering total costs was that 'he who benefits should pay'. But this is a value judgment which may not be generally acceptable. Users of certain public transport services may include many low income earners in whose favour it is desired to effect an income redistribution. It is on such grounds that education, parks, libraries and medical facilities are provided below cost to the users. But the subsidy on these items also implies the imposition of a system of values, e.g. that everyone should have free access to medical services, etc., together with the expectation that many potential beneficiaries would not secure adequate services if the service were not provided centrally under subsidy. It is not obvious that either of these premises is relevant to the case of transport.

A second argument for subsidization may be quite consistent with the 'he who benefits should pay' principle. Under a normal market mechanism a transport agency is only able to secure payment by those who benefit directly by use of the service provided. But, it is argued, some of the benefits that arise from the provision of a transport service though indirect are nevertheless substantial. Consider the following cases:

(i) The provision of a subsidized transport service may stimulate employment in the depressed areas. The element of subsidy may have only just made it possible for certain undertakings to operate. Hence they cannot be charged any more. But in fact the social cost of their output is less than the private cost as the labour which they employ was not previously producing anything and thus has no real opportunity cost.[1] Benefits accrue to the workers in so far as their income from employment is greater than their unemployment assistance and to the taxpayer in so far as the burden of assistance is reduced.

(ii) The operation of a railway line, by keeping a certain amount of traffic off the roads, may help to limit congestion. This has a two-fold effect; firstly, it keeps down the private costs of the road operator and, secondly, it saves the community the cost of increasing road capacity. Some benefits therefore accrue, without cost, to the road operator and to the taxpayer.

In both of these cases there is no direct link between the provider of the service under consideration and those who benefit so that there is no question of being able to devise a pricing system whereby the operator can appropriate directly to himself any part of the gains discussed. If the beneficiaries are to pay, this can only be achieved by a subsidy financed by them.

Thirdly, subsidization may be a method of implementing other aspects of public policy. If it is deemed desirable to encourage a particular geographical distribution of industry, or to maintain the present population distribution, State subsidy of transport may prove an effective, though possibly costly, instrument of policy.

Finally, in situations in which a transport operator is unable to cover his total money costs by any charging scheme, the decision may be made that certain costs should be disregarded in the interests of securing a maximum use of existing assets. This would, as we have explained previously, carry implications for the distribution of income.

To summarize then. It appears desirable that full costs be recovered from users because this appears equitable, because it is a policy which can be pursued alike by public and private enterprise, making it possible for consumer choice to be based on comparison of cost, and because it offers a rough incentive to control the level of costs. On the other hand we have been able to envisage situations where to follow the rule slavishly would be to accentuate rather than eliminate distortions in the comparison between agencies, or where it would perpetuate social problems which a more flexible approach might help to ameliorate. Therefore, in going on to consider possible pricing techniques, we must set out the implications and implicit value premisses attached to particular techniques in the knowledge

---

[1] This assumes that labour is currently immobile in the geographical sense.

that different basic situations may call for entirely different pricing policies.

How should operators recover their costs?

In the last chapter we made a distinction between direct and indirect cost on the ground of escapability. Direct costs were defined as those costs escapable with respect to the unit of decision within a specified time period. Indirect costs were the residue which could not be specifically attributed in this way. Thus the essence of indirect costs is that there is no obviously correct way of recovering them.

There are, however, varying, degrees of escapability or directness. Some costs are caused by the existence of a particular route, though it is possible to allocate them between services on that route; some, though still not imputable to particular individual passengers or freight consignments, may be imputable to a particular lorry, bus or train load. We have previously argued that at these various levels of decision opportunity costs are involved in each case. As the opportunity cost indicates the value of the alternative foregone it would appear sensible as a general rule to insist that, whatever the unit of service provided, traffic should not be carried for a revenue less than the direct cost incurred.

It might be thought that we could rely on operators to ensure that this rule was observed. But there are two circumstances in which operators have frequently offered their services below direct cost.

(i) When they do not realize what their direct costs are. A direct cost rule is firstly therefore an injunction to find out about costs.

(ii) Where they feel obliged to provide some services below direct cost as a social duty.

It is clearly not sensible to attempt to cover direct cost and fail for lack of information; it is not so clear, however, why operators should not be permitted to indulge their social conscience by knowingly offering their services below direct cost. After all, in the last section we suggested several reasons for condoning a failure to cover total cost.

The justifications adduced for a failure to cover total costs all extended beyond the immediate sphere of railway operations to wider issues such as income distribution, or social policy, which are clearly matters of Government responsibility. The arguments supported a possible State subsidy. For a transport operator to supply services below cost of his own accord is a different matter. In the first place, if the operator is a public corporation it may well fall into an overall deficit which would have to be financed from central funds. A public corporation which, without the consent of the Government, puts itself in this position is usurping powers which rightly rest with the Government.

To the extent that the transport operator is both trying to cover his costs and to act as a public benefactor by providing some facilities at below direct cost, the implication is that the loss incurred on those services must be made good from other traffics. The finance of the operator's public service obligations is thus internal. The objections to such internal cross-subsidization are twofold, namely:

(i) That the burden of providing cheap transport for some users should fall only on other users of the same agency seems arbitrary and inequitable.

(ii) If a competitive situation exists between two operators, only one of which has any public service obligations, business may well be diverted to the operator with higher costs. For example, consider two operators, X and Y, competing in two markets, A and B. We will simplify the situation by assuming that there are no indirect costs and that short run marginal costs are constant. The direct costs for X are £6 in market A and £5 in market B whilst for Y they are £5 in market A and £6 in market B. If both operators offer their services at cost and there are six units of traffic in each market, X will meet all the demand in market B and Y all the demand in market A. The total cost to the community will be £60.

| | Market A | | Market B | |
	Price	Output	Price	Output
Operator X	£6	0	£5	6
Operator Y	£5	6	£6	0

But suppose that X decides that it would be a public service to give consumers in market A freer choice by equalizing charges. He thus reduces his price in A to £5, obtaining half the traffic on which he loses £1 per unit. Now in order to meet these losses in market A he must make a profit in market B. Assuming that prices must be set in units of £1 he can best do this by raising his price to £6. He loses half of this market but makes £1 per unit profit on what is left. The outcome is that both markets are divided equally between the operators, neither has improved his overall profitability, but the total cost to the community has risen to £66 due to the fact that neither operator is concentrating specifically on that traffic for which he has a cost advantage.

	Market A		Market B	
	Price	*Output*	*Price*	*Output*
Operator X	£5	3	£6	3
Operator Y	£5	3	£6	3

Internal cross-subsidization to support prices below direct cost is thus unsatisfactory both on grounds of equity and of economic allocation of traffic.

Having made the general recommendation that price should at least cover direct cost we still have to solve the difficult problem of what to do about indirect cost. A number of alternative techniques could be employed.

(i) *Averaging.* The first possibility is that a surcharge, fixed either in amount or proportion, be added to direct cost. On the basis of estimates of the likely volume of traffic, this surcharge would be set at a level sufficient to ensure that enough revenue is raised to meet indirect costs in total. Since, by definition, indirect costs cannot be meaningfully imputed to particular units of output it seems fair to recover them by saying that each unit of service should make an equal contribution.

Despite the attractive simplicity of this proposal it is not without disadvantages. Where the operator provides a number of different types of service he is forced to 'average' over units which are strictly non-comparable. For instance, the allocation of overheads between passenger miles and ton-miles would be essentially arbitrary, it being impossible to devise a uniquely acceptable way of converting one into the other. But, even with a homogeneous output, it might be objected that averaging would lead to an unduly limited use of the facilities provided by turning away potential customers willing to pay more than the direct cost of the service required but not able to pay the required level of contribution to indirect costs. In fact, by refusing such traffic the average contribution of existing traffic will be increased.

(ii) *The multi-part tariff.* Instead of charging through unit prices it might be possible to divide the price charged into two or more parts. Firstly, there would be a fixed charge payable by all consumers, or a range of output charged at a high price, set so that in total all indirect costs are met. Secondly, there would be a charge per unit based on the direct costs of using the service. In this way the average total cost to any consumer would fall the more units of service he purchased, thus encouraging intensive use of the facilities provided. This does not completely eliminate the problem of under-utilization of assets though. Although marginal consumption by those who have paid the fixed charge is encouraged by the fact that the price at the margin is related to direct cost only, nevertheless some potential consumers who would be willing to meet this marginal charge will not be eligible to do so as they have been deterred altogether by the lump sum payment or first part of the tariff. Some who would have purchased the service even at an average cost price will not choose to purchase at all under a two-part or multi-part tariff because their total consumption potential is either small or unpredictable.

Adoption of this type of pricing policy is only practicable when users are identifiable and the product is non-transferable. It is therefore in principle applicable both to passenger and freight transport and has been used for some rail freight traffics since 1957. For the rail passenger, however, direct costs form a very small part of the total costs. Therefore the first part charge would have to be very large if costs were to be covered. The effect of this would almost certainly be to magnify the deterrent effect of this first charge and lead to a very poor rail utilization.

(iii) *Discrimination.* In order to permit marginal consumption at direct cost and yet recover indirect costs without incurring the deterrent effect of the first part of a two-part tariff it is necessary to abandon the notion of uniform prices altogether and to discriminate not only between different blocks of output but also between customers themselves. In other words price would be based in the first place on direct cost and indirect costs would be allocated between traffic according to 'what the traffic will bear'. Consumers who are marginal would thus make little or no contribution to overheads whilst those who are intra-marginal will be squeezed for as much as possible. This is certainly compatible with the principle that he who benefits should pay as the amount that can be squeezed from the consumer cannot be more than the net advantage which he considers the particular form of transport concerned to have over others. Discrimination, in this context then, is simply a method of allocating indirect costs according to the magnitude of the net benefits resulting from the service. 'What the traffic will bear' in this sense is thus not directly dependent on the value of the consignment transported but on the cost of its carriage by other agencies and the magnitude of the benefits accruing to the consignor.

Discriminatory charging according to what the traffic will bear thus combines the following advantages.

(i) Anyone who values a service more highly than the direct cost of its provision is permitted to use it, thus encouraging intensive utilization of capacity.

(ii) It enables indirect cost to be recovered from beneficiaries without resort either to general taxation or to any charge which will distort choice at the margin.

(iii) It enables indirect costs to be distributed amongst beneficiaries roughly according to the magnitude of the net benefit.

THE IMPORTANCE OF COMPARABILITY

Despite the apparent superiority of the principle of charging according to 'what the traffic will bear' we still cannot conclude that it should always be used by transport operators. Our main concern is to secure that competition between operators should be so regulated that any given output of transport services be provided at minimum cost. But if methods of charging are not uniform distortions can still occur. Suppose, for example, that Operator X is charging according to direct cost plus 'what the traffic will bear' while his competitor, Operator Y, is forced for some reason to average his indirect costs over all units of traffic. If some part of X's demand is highly inelastic, because of substantial technical advantage over Y for some types of traffic, he may be able to effectively ignore the competition of Y by covering the whole of his indirect costs through contributions from consumers in this sector. He will then be free to charge at only direct cost in markets where competition from Y is more rigorous. It would then be possible for X to undercut Y even though Y had lower direct, and possibly total, costs in this sector merely because Y was being forced to attempt to recoup his indirect costs from this traffic while X is not. Thus, if one operator is unable to use our discriminatory charging scheme it may be in the interests of a rational allocation of traffic to prevent competitors from discriminating.

If a single pricing rule was to be introduced in transport a system based on direct cost plus 'what the traffic will bear' would appear to be the strongest candidate for the reasons already expounded. There is one reservation, however, which ought to be mentioned here. We have assumed that by use of this charging technique not only will traffic be distributed currently in a desirable manner but, by the success or failure of an operator to cover indirect costs over particular ranges, some indication will be given of the total capacity requirements. It may well be, however, that in some cases while two competitors are providing alternative services neither can recover indirect costs. In such a situation (which will be spelt out in detail in Chapter 12) a rational outcome can only ensue if the operator with the higher total cost when carrying all the traffic, over a suitable period of time, is eliminated.

SUMMARY

No single pricing technique can be recommended for application in all situations to yield a desirable outcome. The best we can do is to assert the importance of ensuring that competing agencies adopt policies which are comparable in operation, yet which do not involve any immediate distortion towards high real cost allocations of traffic or imply any value judgment concerning the redistribution of income involved which would not be generally acceptable when stated explicitly.

CHAPTER FOUR

The Investment Decision

The annual expenditure on fixed assets for the production of transport services in this country amounts to well over a thousand million pounds. In 1961, direct consumers expenditure on cars and motor cycles alone amounted to £521 million. In the same year industrial gross fixed capital formation in various forms of inland transport is estimated to have been £684 million, or about 15 per cent of total gross fixed capital formation. (See Table 2.)

TABLE 2

Industrial Gross Fixed Capital Formation in Inland Transport, 1961

	£ million
Railways	154
Public road passenger transport	43
Roads	106
Harbours, docks and canals	23
Air transport	24
Other vehicle purchases (all industries)	334
Total capital formation, inland transport	684
Total capital formation, all industries	4,465

Source: National Income and Expenditure 1962. Table 58.
Notes: Half of the total capital formation in air transport has been imputed to inland transport.

The table does not include investment in buildings and plant, other than vehicles, used for transport purposes by road hauliers or other firms.

Nearly half of this investment in transport consists of the purchase of road vehicles by private firms and is a subject of public control only to the same extent and for the same reasons that the Government seeks to manipulate the general level of industrial investment. This still leaves a half of the total, however, which is either undertaken directly by a central or local Government agency or is controlled indirectly by means of ministerial powers over the investment of the nationalized corporations.

If, in the last chapter, we had been able to state a single indisputable pricing rule for all forms of transport service we might have been able to deduce from it a similarly universal investment criterion. As it is, we have no choice but to treat the problem in a more pragmatic fashion.

We can begin by recognizing two levels of decision concerning public investment in transport, namely:

(i) The allocation of public investment funds between sectors,
(ii) The choice of particular schemes within the limitations set by this overall allocation.

There is no doubt that the two parts of the problem overlap considerably; on the one hand the investment projects implemented are limited by the availability of finance, whilst on the other the allocation of funds to any particular sector may well depend on what schemes are proposed within that sector and how well they are advocated. If the economy was very simple and the benefits available from any public investment project could be adequately assessed in terms of its commercial profitability then only one problem would exist, namely, how to choose particular projects irrespective of the sectors in which they occurred. But public investment is not of this simple nature. In many cases it has been decided, either on grounds of policy or as a matter of practicability, not to charge an economic price for the product of sector, thus making it illogical to allocate investment funds simply on the basis of expected commercial profitability. We shall begin therefore by considering how investment funds are distributed between sectors.

THE ALLOCATION OF
PUBLIC INVESTMENT FUNDS BETWEEN SECTORS

Little evidence is available concerning the distribution of investment funds between sectors.[1] Even a White Paper on the subject, published in November 1960, goes no further than to indicate that on occasions it is necessary to use Government control over investment in the public sector as a means of controlling the general level of economic activity. It tells us, in Paragraph 11, 'Each summer the Government considers

[1] When the Transport (Borrowing Powers) Bill was being discussed in the House of Commons in 1955, Mr Holt, the Liberal Member for Bolton West, asked how this allocation was performed and received from the Parliamentary Secretary to the Treasury, Mr Molson, the following reply: 'I would say that we do our best but we do not do it in quite the same way as the Socialists did it. Before deciding how much was to be spent on the roads and how much on the railways the matter was naturally considered by the departments concerned, particularly by the Treasury which is primarily concerned with keeping the economy of the country in equilibrium.' *Hansard House of Commons Debates*, 1954–5, Vol. 536, Col. 1850–51. The most that can be fairly read into this answer is the suggestion that the overall watching brief of the Treasury would be used to ensure that the investment plans of the nationalized industries did not constitute a seriously inflationary pressure.

the proposals for expenditure on public investment in the following financial year put forward by the authorities concerned. . . . The various programmes are considered in relation to the total outlay proposed and against the background of the economic situation. Levels of expenditure are approved for public investment as a whole and for each programme. This review is the focal point in the control of public investment.'[1]

In the case of the B.T.C., for instance, both the size of grant and its allocation between major uses would have been determined by negotiation between the Commission on the one side and the Ministry of Transport and the Treasury on the other, taking as a starting point the B.T.C.'s original estimate of its requirement. This original estimate submitted by the B.T.C. would itself be the result of negotiation and compromise, the original claims of the operating units having been paired down by the B.T.C. to a total claim which it thought the Government might reasonably be hoped to accept. Under the 1962 Transport Act it is implied that this first stage will be undertaken within the framework of the Nationalized Transport Advisory Council.

Not only do the nationalized transport agencies compete with each other for investment funds, they also must compete with other nationalized industries and with the various non-commercial services operated by the Government. With such diversity of organization and type of asset involved it seems unlikely that there is any mechanism which ensures the comparability of the criteria used in the various sectors as the basis of their claims. It is therefore inevitable that this allocation is largely determined by political judgment.

INVESTMENT DECISIONS WITHIN THE TRANSPORT SECTOR

The absence of any objective principles for the distribution of investment funds between sectors does not detract from the desirability of allocating funds sensibly within the transport sector itself. The rest of this chapter will therefore be concerned primarily with this problem.

First let us look at what appears to be the general principles used to decide how much investment is made in any privately owned concern. There appear to be two respects in which considerations of profitability are of great significance, namely:

(a) The past profitability of the enterprise.

---

[1] *Public Investment in Great Britain*, Cmd. 1,203, November 1960.

(b) The future prospects of the enterprise and particularly the profitability of the proposed project.

The role of past profitability

Past profitability is important for two reasons. In the first place, with an ever increasing proportion of investment being financed from retained profits the past profitability of the firm provides the source of the funds invested, and, if there are any reasons why it is preferred not to raise capital on the market, it may be a pre-requisite of further investment. On the other hand, when funds are to be raised on the market it is usually impossible to give to the investor precise details of the project concerned. Hence the investor is forced to lend partly on trust; the guide to the trustworthiness of the concern being the success which it has had in the past in making profitable investments.

Looking strictly from the economist's point of view we might argue that the past is irrelevant, that bygones are bygones and that the only important consideration ought to be the use to which the newly acquired funds are to be put. In fact uncertainty about the future usually necessitates consideration of past profitability. The nature of this consideration is most pertinent when we try to apply it to transport. The investor is interested in making profits. His assessment of the worthiness of any undertaking rests on its degree of success in the past when we presume that a general record of profitability (though not necessarily maximum profits in any particular year) was one of its main concerns. If there is any reason why the undertaking has not chosen, or has not been able, to concentrate on profitability in the past then its success and its future prospects should not necessarily be measured by this yardstick.

Depreciation and replacement

Having made these reservations about past profitability we shall now concentrate on the prospects of the project under consideration. The first problem that arises in this respect is to decide what exactly we mean by investment. In particular we may be worried about the distinction between new, or net investment, which is by definition that part of investment which constitutes an addition to our capital stock, and replacement, which is that part of gross investment necessary to maintain intact the existing capital stock. The accountant's preoccupation with depreciation, which makes provision for replacement, may give the impression that replacement is imperative and of a different order from new investment. For the economist there is no distinction in principle between new investment and replacement. Even if we considered a world in which there was no technical progress we should only wish to replace a machine with another one like it so long as the demand for its products had not fallen off and it was still the most profitable use of the available funds. In a world embracing technical progress we rarely replace like with like and the distinction between that part of an investment which represents replacement and that part which represents 'betterment' may be difficult, if not impossible, to make. This distinction may be important to the accountant in assessing tax liability or simply, as in the case of the B.T.C., in deciding whether the cost should be imputed to current or to capital account, but it is not significant to the economist in deciding whether a particular scheme ought to be implemented or not.

For any asset which is not physically everlasting, a decision will eventually have to be made concerning how, if at all, it is to be replaced. Obviously when the asset breaks up the decision can be postponed no longer; there is no reason, however, why it should not have been replaced long before this if a different machine, or a newer machine, would have proved more profitable in operation. In periods of rapid technological change it may be desirable to scrap assets long before the end of their physical life and even before the end of the shorter replacement period sometimes used by the accountant in an attempt to allow for obsolescence.

The fact that replacement should not occur automatically, even when a reserve has been built up to maintain capital intact, does not make provision for depreciation redundant. No one will be inclined to invest in an asset unless he can expect sufficient earnings from it both to maintain his capital intact and to yield a satisfactory rate of profit. Therefore depreciation is, *ex ante*, a proper long run cost of the asset which the investor hopes to cover. Once the investment is made, however, the relevant short run costs of output from the asset are the costs involved in its use, which, in real terms, are the opportunities lost by its use. If present use of the asset involves wear and tear then its possible use in the future is reduced. Depreciation to cover this lost opportunity is clearly part of the true 'user cost'. On the other hand if the asset becomes redundant after a certain period of time irrespective of the amount of use to which it has been put during its life then use of the asset does not involve the loss of any opportunity and hence cannot be included as part of the 'user cost'. So long as the revenue earned by the asset is greater than the user cost as we have defined it then it is advantageous to keep an existing asset in use. But for new investment to be worth while not only must expected earnings be sufficient to maintain capital intact but also to produce a certain minimum rate of return over and above this.

The rate of return

The minimum rate of return on capital which will persuade the investor to make an investment represents the degree to which present income is preferred to postponed income, given the degree of uncertainty attached to the future income forthcoming from this particular investment. Hence the present value of every £ of income will be greater the sooner the income is realized, and the lower is the rate at which the future is being discounted. This rate of discount will be high if risks are great. The value attributed to the investment at the present will thus be not simply the sum of all the future annual incomes deriving from it but the sum of the annual amounts discounted at the appropriate rate for the appropriate period of time.

The income produced by an investment is unlikely to remain constant each year until the asset is scrapped; for instance, it may well decline slowly from the peak level of annual return due to the gradual physical deterioration of the asset or due to the fact that it is being made obsolete by newer techniques. There may even be an outlay period which is completely sterile, involving no return at all; if this is so, all the eventual returns should be discounted over the period from the original expenditure and not from the first returns. Moreover, this distribution of returns through time will vary from investment to investment. Consequently with two fluctuating income streams the relative value of the streams at present may depend crucially on the rate at which they are discounted. The greater the rate of discount the less attractive will the streams containing the more distant returns appear to be.

In principle we must be continually questioning whether we would be better off if the existing physical assets were replaced. One method of making this decision is to compare the present value of the stream of gross after tax revenues from the old asset (receipts less working costs), discounted at a suitable rate, with the net revenues (again after tax) from the new asset after deduction of interest and depreciation charges but also discounted over time at the same rate. If the rate of discount represents the expected rate of interest[1] and the second sum is greater than the first then this indicates that it is worth while undertaking the replacement of the asset.[2] Alternatively we may use the discounted cash flow method which shows the rate of

---

[1] Adjusted to take account of variations in risk if necessary. See Bierman's *Managerial Accounting*.

[2] On the assumption that the supply of funds to the firm is infinitely elastic at the interest rate used in the calculation. If this were not so the project showing the greatest rate of surplus, i.e. the greatest present value per £ invested, would be chosen.

return which the expected income stream represents on investment whilst automatically making allowance for depreciation in accordance with the time pattern of the income streams.[1] In this case the investment whose cash flow table showed the greatest internal rate of return would be the best investment to make.[2]

The simple scheme, which we have outlined so far, can be applied almost without modification in many industries. It depends, however, on two assumptions which frequently do not hold in the transport industry. In the first place, it assumes that there is an increment of output resulting directly from our increment of capital which can be measured and expressed as the return on capital. In railways there are many forms of investment which do not appear to have any direct yield but which nevertheless are necessary in the long run to keep the whole system ticking over. We shall call this the problem of indivisibilities of service.

Secondly, it assumes that the invested capital yields a product which is sold on a market and hence has a market price. But this is not the case with investment in roads. There is no direct market price for use of the roads and hence the question of the rate of profit cannot arise in the way in which we have discussed it in our simple scheme. We shall call this the problem of non-monetary benefits.

Indivisibilities of service
The first thing to notice about indivisibility is that it is not a problem specific to transport. In manufacturing industry there are many services which do not yield a discernible output for each unit of input yet which are still considered essential factors in the productive process. Advertising and internal information services are examples. However detailed the records and accounts of a company may be it is impossible to express with any precision the profitability of the central service departments. Hence the decisions as to the size of these departments will tend to result from a judgment rather than a calculation. A great deal of the physical capital of the railways is of this kind. Signalling is obviously necessary but it may not be at all easy to assess a rate of return on a new signalling system to replace that currently in use.

There is one respect in which investment even in signalling yields a measurable benefit. Though it may be impossible to obtain an

[1] For a further discussion of these techniques see J. Fred Weston, *Managerial Finance*, pp. 122–30.
[2] These two alternative methods may give different results where the interest rate used is not equal to the internal rate of return.

E

estimate of increased revenue resulting from the faster travel obtained by signalling improvements, the capital cost may itself be partly offset by reduced running cost which can be directly measured. Indeed, it is one of the dangers of investment in railways that there may be more emphasis on projects which promise reductions in direct cost than in those which promise improvements in gross revenue simply because they are more certain and more easily measurable.

Even in cases where increments of capital have a certain and measurable effect on receipts we may still be faced with difficulties. Though the increment may yield improvements in net revenue which may be expressed as a rate of return on the capital involved, the investment may be inseparable from a line or sector which is in total unprofitable. This is a situation unlikely to arise in most industries where the whole emphasis would be to weed out those unprofitable parts of its operations but the historical peculiarities of transport, not only in Great Britain but in other countries as well, have resulted in the continued operation of such sections. Because investment in transport tends to be very specific, once embodied in the equipment required it is very inflexible. Hence if a railway line is closed the transfer value of the assets is fairly low (especially when railways as a whole are contracting and it is not possible to transfer the assets from one part of the system to another). Thus there seems to be a dual problem involved here. It is important to secure a satisfactory rate of return on the capital investment but it is also imperative to ensure that good money is not being thrown after bad and that the closure of the line or section will not render the capital redundant in the near future.

There has been some disagreement in academic circles as to whether or not these two requirements could be combined in a single investment criterion. Mr D. L. Munby has argued that no single test would be a sufficient guide in all circumstances[1] whereas Dr M. E. Beesley has produced a single criterion.[2] This disagreement is more apparent than real however. Mr Munby, as part of his conclusion, states that to decide on the exact nature of the problem and set out any restrictive assumptions which must not be contravened by the solution is a very important pre-requisite for any test. Dr Beesley accepts, and makes deductions from, a very rigid set of assumptions. Railway management is assumed to be bent on maximizing

[1] D. L. Munby, 'Investment in Road and Rail Transport', *Journal of the Institute of Transport*, Vol. 29, No. 3, March 1962.
[2] M. E. Beesley, 'Financial Criteria for Rail Investment', *Bulletin of the Oxford Institute of Statistics*, Vol. 24, No. 1, February 1962.

the present value of its assets, and to have perfect knowledge of all the possible improvements, capital supplies are assumed to be limited by an arbitrarily fixed annual budget and there are no restrictions on the freedom of the railways to set their own prices or to close any lines which are deemed unprofitable. By using a time discounting procedure on both costs and receipts, Dr Beesley produces a single investment criterion which is quite consistent with his premisses.

The list of assumptions is formidable and the main conclusions that one can reach from Dr Beesley's rigorous analysis is that the railways must do a great deal more to identify the costs and revenues appropriate to a project before such a criterion could be applied with any precision. They would need detailed information concerning interdependence of services; for instance, the extent to which the withdrawal of a particular feeder service would affect traffic on the rest of the railway system. Above all, it is clear that the assumptions made about the future availability of capital funds are crucial in determining whether a line should be closed or not, even though the task of the railways is to maximize the present value of its assets. Thus, if the railways are to make sensible commercial decisions it is important that they be given some long term assurance concerning the availability of funds.

Non-monetary benefits
We can demonstrate the problems involved by the non-monetary benefits resulting from an investment by a series of diagrams. Let us start off with the simplest of economic pictures, that of price formation where supply and demand curves for a product intersect (see Figure 3). SS is the supply curve, DD is the demand curve intersecting it at X, so that the price is OP and the quantity sold OQ. The amount of money paid to the producer for this price-output combination is OPXQ. But this is not a full measure of the benefits ensuing from the production of OQ. Some consumers would have been willing to pay more than the market price OP for their unit of product so that the total benefits are more than OPXQ. The extra benefit has traditionally been called the 'consumers' surplus' and is in our diagram equal to the area PDX. Clearly, given DD, the lower the price the greater the consumers' surplus. Now so long as OPXQ is greater than the total cost of producing the output OQ (and in the long run it must be, unless all producers have the same costs, or at least some of the producers will go bankrupt) it can be shown that the producers are obtaining some surplus over their costs of production so that producers' surplus (or profit) is obtained at the same

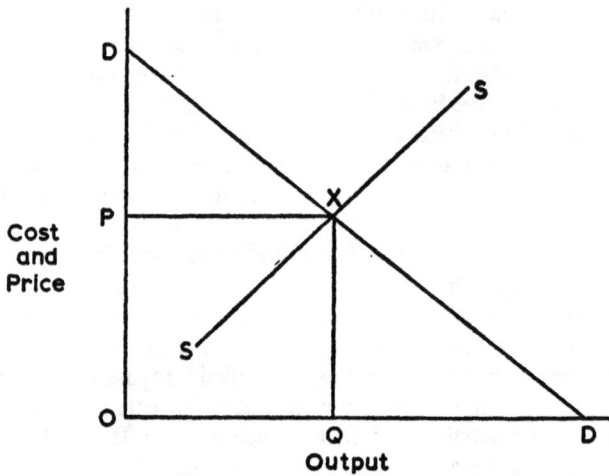

Figure 3

time as consumers' surplus. Our profitability criterion measures the rate at which producers' surplus occurs as the result of capital investment but ignores consumers' surplus.

Now let us consider a product or service with price zero, but the demand still DD' (Figure 4). There is no price so that there is now no revenue accruing to the producer. Consumers' surplus amounts to the whole of the triangle DOD', below the demand curve (which we have taken for convenience to meet both axes) and is at a maximum. Thus, in these circumstances, the maximum rate of consumers' surplus to capital invested will be achieved with zero price.

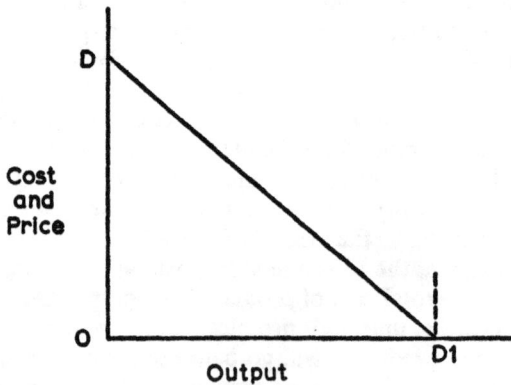

Figure 4

Even if we assume that there are other costs (such as maintenance) which are *not* related to use of the asset then the maximum ratio of consumers' surplus to capital will be achieved by price=0 because then the consumers' surplus is greatest while cost is constant irrespective of the use to which the asset is put. This can be demonstrated (see Figure 5) by a rectangular hyperbola CC' (such that the

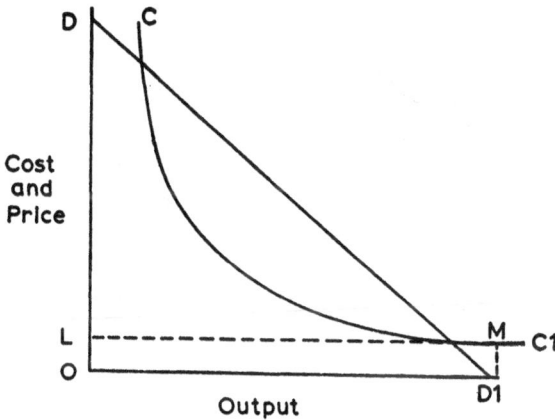

Figure 5

rectangle formed with the axes is the same for any point on CC'). The total cost is then ODML and the maximum possible consumer surplus is DOD'. The rate of benefit can then be expressed as

$$\left(\frac{DOD', -OD'ML}{\text{capital invested}}\right) \times 100 \text{ per cent}$$

The cost to the community in this case is constant irrespective of use of asset so that the maximum rate of benefit on capital cost will be obtained with price=0 and it would appear to be sensible to invest in those lines where the rate of social benefit on capital is at a maximum.

A more complicated analysis is necessary where variable costs arise. There are then three surpluses which could be maximized. Firstly, we could concentrate on the maximization of producers' surplus, or profit, which is achieved by extending output to a level at which the marginal costs is equal to marginal revenue (see Figure 6).

Secondly, the maximization of consumers' surplus would be achieved by zero price, but only at the expense of subsidy from the producer. Consumer surplus *above cost* would be achieved by

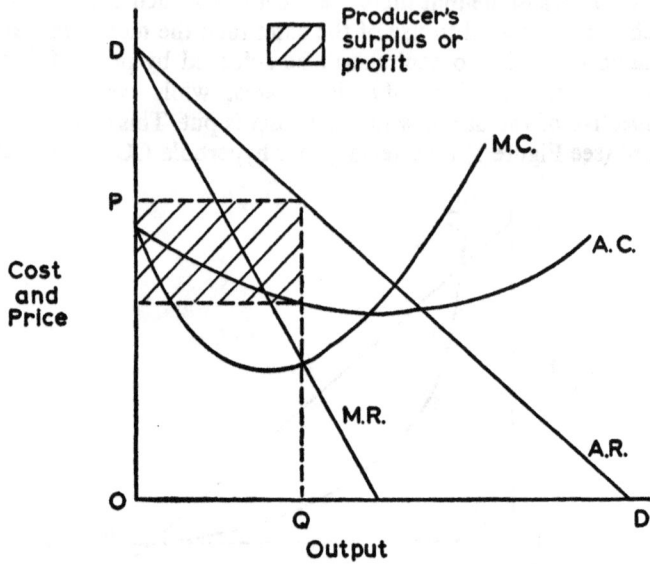

Figure 6

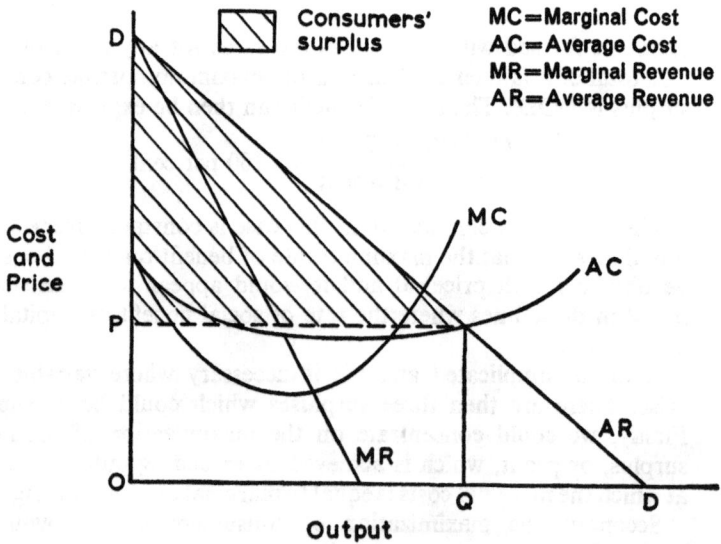

Figure 7

extending output to the level at which price just ceases to exceed average total cost (see Figure 7).

Finally, we might wish simply to maximize the surplus above cost irrespective of its distribution between producers and consumers. This would be achieved by producing to the point where price just ceases to exceed marginal cost (see Figure 8). This total may be called the 'social surplus'[1] and would appear to be the most obvious candidate for maximization in public investment.

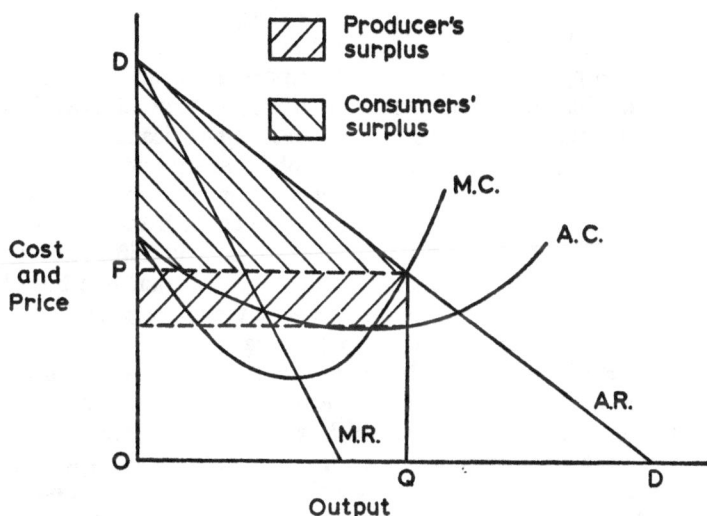

Figure 8

Unfortunately, a number of drawbacks arise.[2]

Firstly we may not be able to measure consumers' surplus in any useful way. Some of the savings due to an investment may be obvious reductions in direct cost. Straightening a road may reduce operator's petrol consumption and by speeding up transit may reduce labour costs. Each of these has a market value and so apparently could be aggregated. There may, however, be savings of leisure time involved which have no market price so that aggregation is only possible by imputing a value. Benefits may accrue to operators on other roads

[1] If average cost is greater than marginal cost at the point where marginal cost equals price the social surplus will consist of consumers' surplus less producers' deficit. Otherwise social surplus equals producers' surplus plus consumers' surplus.

[2] For a more general analysis of consumers' surplus see I. M. D. Little, *Critique of Welfare Economics*.

where congestion is reduced as a result of the original improvement. The ripples from our original disturbance may be very widespread and difficult to trace; when traced they may defy measurement. In the chapter on roads we shall consider in greater detail the practical difficulties which arise when one tries to measure the surpluses involved in the M1 project. Aggregations of this kind using imputed values necessarily involve value judgments and therefore the calculation will not be unique or unobjectionable.

Secondly, we must note that the magnitude of the consumers' surplus is not independent of the price charged either for use of the asset in question or for the use of other facilities complementary to it or competing with it. This is a most important consideration in view of the fact that the social surplus criterion has been advocated particularly for public investments where it is impracticable to charge a price for use of an asset, either on physical grounds or because competing facilities are not being charged for at an economic level. Thus, for instance, the apparent benefit to ensue from the Victoria line extension of the London Underground will differ not only according to the level of fares to be charged on that line but also according to the fares charged on the rest of the underground railway and according to the price effectively charged for use of competing road routes.[1] If there could be no question of changing any of these associated prices, it would make sense simply to ignore this matter and implement those schemes which would yield the greatest social benefit in the given circumstances. If possible, however, it would be desirable first to make such price adjustments as were necessary to secure the most economic use of existing assets and only to consider the social surplus produced by new investments in this context.

Thirdly, even if we can measure the surplus satisfactorily, how does it compare with the profitability method of choosing between projects? It is clear that the rate of return expressed in total surplus terms will be greater than that expressed in terms of commercial profitability in all cases where demand is less than perfectly elastic. Hence it will not be justifiable to compare the rates of return for two projects calculated in different ways. Nor will there be any single conversion factor. The effect of a cost-reducing investment must always be to increase the surplus over costs, but it will not necessarily increase profits to the same extent in all market situations. We cannot therefore make any indirect comparison between the profitability of investment in surplus and commercial terms.

[1] See C. D. Foster and M. E. Beesley, 'Estimating the social benefit of constructing an underground railway in London', *Journal of the Royal Statistical Society*, Series A, Vol. 126, Part I, 1963.

Comparability of rates

The problem of comparability of rates of return computed on different bases is not confined to the distinction between the surplus criterion and the commercial criterion. It is commonly argued that you can, and must, compare rates of return in different sectors of the economy and in particular that the nationalized industries must not be allowed to obtain their capital 'on the cheap'. But there is a dangerous fallacy involved in this argument. It is valid only in so far as there is no element of subsidy involved in the pricing and purchasing policies of the industries concerned. In this respect we would define 'subsidy' not only as a direct grant to an individual to set against the market price of a commodity, but to include any situation where the market price is administered at a lower level than that at which it would be set in normal commercial conditions. Hence, we would not define as subsidy the restraint on a monopolist who does not exploit his position to the full for fear of the long-term effects on his position whereas we would define as subsidy any political imposition which prevents a firm or industry from acting within the normal limits of commercial freedom. Both in road passenger transport and on the railways there has been a great deal of cross-subsidization of unremunerative routes from the profits reaped elsewhere with the result that the industries as a whole appear to be unprofitable in situations where there are individual projects available which, if separated from the ethos of restriction surrounding the industry, might appear to be very desirable and profitable schemes, both for the industry and for the consumer.

External economies

There is a school of thought which argues that investment in roads produces economies in the form of reduced transport costs which cannot in principle be recouped from road users. It is argued that these are the 'external economies' of the provision of roads and because they are so much greater in road transport than in other spheres the investment in roads should be treated as a special case to which no formal criterion need be applied.

But the term 'external economies' is frequently used in such a vague manner that we must subject this argument to a close scrutiny. Professor Meade has given probably the most concise formal definition of an external economy as existing whenever the output of a firm depends not only on the factors of production directly employed by it but also on the output and factor use of other firms.[1]

[1] See J. E. Meade, 'External Economies and Diseconomies in a Competitive Situation', *Economic Journal*, Vol. 62, No. 1, March 1952.

Thus the economy of scale in this sense (which may be called the 'technological' external economy) is not involved in the output but in the cheapening effect of the firm's policy on the inputs used in common by several firms or industries.[1] A heavy road building plan may possibly enable the civil engineering industry to achieve cost reductions but there is no sign that this effect is particularly significant in the transport sector.

More usually, however, the 'external economies' argument is taken to include cases where reduction of the direct costs of production of transport is passed on completely to users of the service.[2] The 'external economies' argument for subsidized investment in transport thus resolves itself into the assertion that it is not always possible to relate the profitability of the investment to the benefits provided for its users. This could apply to rail as well as road transport. Scitovsky discerns three conditions which would justify this assertion.[3] The first concerns the use of indivisible factors where price discrimination is not practised. This case certainly occurs both in road and rail transport as a consequence of the large element of joint overhead costs. But it has been artificially accentuated in this sector in the past by the prohibition of discriminatory rail charges and the absence of any attempt to levy charges for road use directly related to cost. We may therefore doubt whether this argument is necessarily of greater relevance to transport than to several other industries where joint overhead costs are important.

Secondly, Scitovsky refers to the inapplicability of general equilibrium theory to problems of investment. He argues that profit maximizing behaviour is compatible with a social optimum situation only in a perfectly competitive system with a satisfactory income distribution. Within this framework profits are a sign of disequilibrium. The elimination of this disequilibrium by the process of investment is an essential part of the market mechanism. But we can conclude that the amount of investment which is profitable is equal to the amount socially desirable only on the restrictive assumptions that perfect competition is existent and that similar adjustments can take place freely in all industries. In practice, imperfections of competition and unadjusted disequilibria certainly occur in the transport sector. But there is no reason to believe that their incidence is of more significance in this sector than in many others so that the argument cannot stand as a justification of the special treatment of transport alone.

[1] Meade, *op. cit.*, refers to this as 'the creation of atmosphere'.
[2] Meade calls this an economy of 'unpaid factors of production'.
[3] T. de Scitovsky, 'Two Concepts of External Economies', *Journal of Political Economy*, Vol. 62, No. 2, April 1954.

Thirdly, a divergence between the profitability of an investment and its desirability from the community's point of view may arise through international effects. Taking a purely national viewpoint the benefits to be considered are those accruing internally. Any benefit accruing externally must be treated as lost. Now consider what happens if new transport facilities are provided unprofitably in order to reduce transport costs. If the elasticity of demand for our exports is greater than unity our export earnings will rise; if demand is inelastic export earnings will fall. This would probably have repercussions on imports, which would affect the internal employment structure, and so on. The size and nature of the ultimate changes in national welfare will depend not only on the various elasticities of supply and demand for the commodities traded but also on a host of political imponderables.

How should these facts be interpreted? Let us admit the possibility that there will be occasions when an increase in international trade is desirable from a purely national point of view and assume that the Government is wise enough to recognize such occasions when they occur. Both exports and imports have to be transported and the provision of transport services below cost will automatically stimulate international trade to a certain extent. But so, of course, would any other form of subsidy on international trade. Indeed, a more discriminating direct subsidy would probably be more effective.

Thus the benefits often described as the 'external economies' of investment in transport do not differ in principle from the benefits resulting from cost reducing investment in other sectors. This is not to deny that there may be situations in which the provision of cheap transport facilities may be deemed socially desirable. For instance, where there is local unemployment new firms may be induced to set up on account of subsidized transport facilities. Inducement by transport subsidy is not without disadvantages of course. Firstly, it may be indiscriminate and wasteful, offering windfall gains to those who need no such inducement as well as to those who do. Secondly, in so far as transport prices are raised elsewhere to finance this subsidy, the allocation of traffic between agencies may be distorted in those other localities. Thirdly, the method is inflexible. In the long run regional aid is only economically sensible if it permits an area to develop its productive potential; if an area proves to be lacking in such potential then continued support is not desirable. Investment in transport infrastructure is an irrevocable commitment to a particular location whereas other forms of inducement might be discontinued.

Thus, though some interpretations of the 'external economies' argument may have valid application to transport, prodigal investment in the sector on account of some special facility for producing benefits of this kind is not justified.

SUMMARY

Within the private sector investment decisions in transport present no special difficulty. In the public sector, however, numerous complications arise. There is no clear criterion for the allocation of resources between transport and other sectors. The absence of an economic pricing system for roads and the indivisibilities and interdependence of railway services, some of which are currently unprofitable, preclude any simple criteria for the individual agencies.[1] Finally, wider social implications may even necessitate investments which would fail to meet the normal requirements of the agency concerned.

[1] Investment criteria are discussed more fully in C. D. Foster, *The Transport Problem.*

PART II

THE HISTORICAL BACKGROUND

CHAPTER FIVE

The History of Intervention

In this chapter we wish to show how and why State intervention in the transport sector has developed.

THE PRE-RAILWAY AGE

In the Middle Ages Britain possessed probably the worst communications system in Western Europe. Only the Church paid any attention to maintenance of the roads which had been constructed during the Roman occupation; the historian Jusserand records that road repair was considered 'a pious and meritorious work before God of the same sort as visiting the sick and caring for the poor'. Unfortunately religious fervour was unequal to the task of maintaining a viable road system and in the mid-sixteenth century responsibility for road upkeep was transferred to the parishes to be performed by compulsory labour.[1] This grossly inefficient system continued until 1835 when the general principle of highway rates was introduced. During the same period the creation of turnpike trusts did produce some improvements but in general road maintenance was scanty and ineffective.[2] Vehicular road transport was only possible by the broad-wheeled long wagon or the sprung coach and it was not until the nineteenth century that an attempt was made to adapt the roads to the vehicles rather than adapting the vehicles to the atrocious state of the roads.

Due to the poor state of the roads coastal shipping and inland waterway transport flourished. The hazards of floods and falls on the navigable rivers were ameliorated by the construction of cuts and locks in the early eighteenth century and the construction of canals was a very natural transition from the improvement of rivers. Once the Duke of Bridgewater and his engineer Brindley had demonstrated that both the financial and the technical obstacles to canal building could be overcome there followed, in the latter part of the

---

[1] By the Highway Act, 1555.

[2] The first Turnpike Act is dated at 1663. In the early eighteenth century there was considerable popular discontent with the system culminating in a series of 'turnpike riots'. Not until the latter part of the eighteenth century was there any great extension of turnpike roads.

century, a mania of canal construction. Many of these new canals immediately became effective monopolies of bulk inland transportation, limited water supply preventing duplication of canal construction and the state of the roads eliminating the likelihood of serious competition from other agencies. In the absence of any statutory control high rates and large profits were rife, much to the detriment of other traders' interests. This background goes far towards explaining the suspicion with which railway developments were subsequently met as it soon became apparent that, by their technical superiority, they would achieve an effective monopoly of bulk inland transportation similar to that held by the canals.

THE CONTROL OF RAILWAY DEVELOPMENT

The earliest railways were complementary to existing facilities, being private carriageways to carry coal from mines to the nearest canal or river. The rapid transformation from a subsidiary to a predominant form of transport was the result of three separate, but often simultaneous, changes. Firstly, the railway company became a public carrier, though this often only meant that byetraders were allowed to operate on the railway for the payment of tolls. Secondly, the advent of steam locomotion vastly increased the efficiency of rail transport and incidentally made it necessary, for reasons of safety, for the railway company to provide all the motive power. Thirdly, the railways, encouraged by the intransigence of the canal companies in maintaining exorbitantly high rates, began to operate in competition with the canals. The Liverpool and Manchester Railway, for instance, was originally supported by traders in the two cities who wished to break the canal monopoly. The commercial success of this railway set the stage for the frenzied bouts of railway authorization and construction which took place in the 1830s and 1840s.

Railway promotion and the public interest

The formation of a railway company necessitated a Private Act of Parliament. Though this was often made very expensive by canal opposition the resulting acts showed little evidence of conscious Parliamentary direction. Safety provisions and speed restrictions were the natural reaction against the unknown possibilities of the new technique whilst restriction on rates and profits were the purely negative results of previous experience with canal monopolies. Otherwise early development conformed to no overall plan. Lines were constructed to deal with local traffic and provision for through traffic was ignored. There were even three different gauges in use,

4 feet 8½ inches being the rule in the midlands and the north, 7 feet in the west and 5 feet on the Eastern Counties Railway. Clearly some rationalization was required before an effective national railway network could be created.

Parliament was not entirely unaware of this problem. In 1838 it rejected an attempt to place the entire railway system under the supreme control of the Postmaster General, an attempt aimed at preventing wasteful duplication of investment.

By 1844, however, Parliament was so submerged with Private Bills that a Select Committee was set up under the chairmanship of Gladstone to consider the whole business of railway promotion. On the strength of its findings a special railways department was set up at the Board of Trade. This body, known as Dalhousie's Board after its chairman, was envisaged as the instrument of Government supervision of railway creation, and possibly eventually of railway operation. Unfortunately it did not last. It engendered great opposition from railway interests and, despite the obvious thoroughness of its investigations, the Prime Minister, Sir Robert Peel, refused to act on its recommendations. It was therefore disbanded in 1845.

The Select Committee of 1844 also reported that there should be 'more rational progress and less indiscriminate multiplication of lines'. 'Competition,' it stated, 'in areas already provided with lines was probably much more efficient as an instrument of injury to existing companies than as a means of guaranteeing cheapness of travel.'[1] The Railway Regulation Act, 1844, which resulted from the report, was, however, an emasculated version of the Committee's recommendations for though it contained severe threats for new companies, it did not impose any very onerous restrictions on those already in existence. The Act required all companies to provide at least one train a day consisting of third class accommodation, travelling at not less than 12 miles per hour, for which the rate was not to exceed 1d per mile—the famous Parliamentary train. Any new company paying a dividend of more than 10 per cent per annum was required to revise its rates and the Government reserved the right to purchase these new companies at any time after 21 years had elapsed since their creation. Thus the 1844 Act protected the consumer by stipulating certain minimum standards of service and by limiting profits and looked forward to the possibility of much more thorough intervention in the future.

The legislation of the 1830s and 1840s was as notable for what it did *not* attempt as for its positive achievements. In the first place,

[1] Third Report of the Committee on Railways, 1844.

F

though it protected the community against the railways, it did not protect the railways against those sections of the community which were in a position to retard their development. Though the State protected landowners from disproportionate damage to their property and ensured that they received fair compensation for land it did not satisfactorily protect the railways against extortionate claims. The Private Bill procedure could be made extremely expensive by judicious opposition. A Bill could cost anything up to £150,000 in the face of opposition from landowning or canal interests, without any guarantee that it would ultimately be accepted. Despite the fact that existence of a railway would often increase land values, opposition was often conciliated only by the payment of compensation for land far in excess of its current market price.[1] Consequently the costs of railway construction in this country were very high.

Even more important was the failure of the State to take any serious steps to achieve co-ordination either between the various railway constructions or between rail and canal interests. It was not until 1846, for instance, that even a standard railway gauge was prescribed.[2] Early intervention thus sought little more than the protection of the immediate financial interests of traders and of the life and limb of passengers.

The protection of traders' interests was supplemented in the 'fifties by legislation concerning through facilities and the prevention of 'undue preference'. Though the statutory recognition of the Railway Clearing House in 1850 consolidated the mutual arrangements between companies to facilitate through traffic which had developed since the creation of the Clearing House in 1842, control was made more stringent when the provision of reasonable facilities for through traffic was made compulsory by the Railway and Canal Traffic Act, 1854. This act also included the famous 'undue preference clause', which required that where a facility or concession was provided for any one trader, the same conditions must be applied to any other trader who required a similar service. The effect of the subsequent interpretation of this clause in the courts was to rob the railways of a good deal of their commercial freedom and flexibility for no fare or charge could now be quoted without simultaneously

---

[1] For instance, the London and Birmingham Railway had to pay £750,000 for land valued previously at only £250,000. See J. Francis, *A History of the English Railway*, Vol. 1, p. 189.

[2] Even the Gauge Act of 1846 was drawn up so loosely as to permit the G.W.R. to construct a further 750 miles of broad gauge track in the next twenty years. See the Report of the Royal Commission on Railways, 1867, p. 76.

considering its effect on the whole structure of fares and charges. This precedent replaced strict commercial criteria as the basis for railway charges. The form of Government control remained, however, predominantly indirect and commercial in character.

Amalgamation

In a sense, the Governments of the day were justified in refusing to intervene in railway operation by the fact that the railway companies themselves, for commercial motives, began to attack the problems of co-ordination and economy of resources. The chosen weapon for this assault was that of amalgamation of companies. The more ambitious managers, such as George Hudson of the Midland Railway, began to extend their operations in this way in the early 'forties and by the end of the decade both the Midland and the London and North Western railways had obtained through routes from London to the Industrial north and both the Lancashire and York-shire and the Manchester, Sheffield and Lincolnshire had created through routes from the East to West across the Pennines. Such amalgamations had to receive the sanction of Parliament, which was readily given where the creation of through routes was concerned.

Not all amalgamations concerned through routes however. Occasionally the effect was to establish a complete regional monopoly, as in the creation of the North Eastern in 1854 and the Great Eastern in 1862. In contrast, other similar schemes were rejected. In 1854 a Select Committee recommended against second readings for bills which would have amalgamated the Midland with the L.N.W. to produce a monopoly between London and Manchester and the L.S.W. with the London, Brighton and South Coast. Again in 1871–3 a burst of unsuccessful amalgamation proposals followed the L.N.W. approaches to the Lancashire and Yorkshire.

In order to secure a clear declaration of public policy concerning amalgamations a Joint Select Committee was appointed in 1872 to consider the matter. Its conclusions, though realistic, did not really clarify the issue. 'Competition,' it reported, 'exists only to a limited extent and cannot be maintained by legislation.' And again, 'Combination between railway companies is increasing and is likely to increase whether by amalgamation or otherwise.'[1] The Committee suggested that it was impossible at that stage to lay down any general rules determining the limits or character of future amalgamations and the method of dealing with such proposals remained effectively

[1] Report of the Joint Select Committee on the Amalgamation of Railways, 1872.

unchanged.[1] By the turn of the century there were only about a
dozen major railway companies left but Parliament was still sufficiently averse to regional monopoly in 1909 to reject the proposed
amalgamation of Great Northern, Great Central and Great Eastern
railways.

Standards of service

Apart from the 'Parliamentary train' requirement of 1844 there was
little control over the quality of the service provided. Fortunately
considerable improvements were achieved simply due to normal
commercial pressures. The Midland Railway led the way by beginning, in 1872, to provide third class accommodation on all trains
and successively improving its standards. Other railways, however
unwillingly, soon followed suit. Hence competition, in this respect at
least, achieved results which many imagined to be only possible as a
result of State intervention.

Railway charges

There was one outstanding problem in the nineteenth century on
which neither Parliament nor the railways themselves made much
impression. This was the matter of the rationalization of charges.
Private bills of railway promotion usually included a classification of
the traffics expected on the line together with a schedule of maximum
tolls and charges. Each company had its own classification and
schedule with the result that for a through traffic carried over the
lines of several companies, determining a charge was a major
operation. The situation was ameliorated to some extent by the
creation of the Railway Clearing House, which facilitated communication and negotiation between companies and also provided
a new unofficial classification, but not sufficiently to eliminate the
understandable dissatisfaction among traders.

Despite the recommendations of a Royal Commission and two
Select Committees earlier in the century, it was not until the Railway
and Canal Traffic Act of 1888 that an attempt was at last made to
solve the problem. Under the provisions of this act all schedules of
maximum rates were revised by the railway companies and reviewed
by the Board of Trade. However, the introduction of the new rates
early in 1893 caused such uproar amongst aggrieved traders that a
Select Committee recommended that maximum rates be frozen at

[1] Proposals for an extra-Parliamentary control were strongly opposed. Even
the Parliamentary Joint committee appointed to consider amalgamations' Bills
under the provisions of the Regulation of Railways Act, 1873, did not meet again
after that year.

their original levels. The Railway and Canal Traffic Act which followed in 1894 extended this restriction to prevent any increase of rail rates even within the statutory maxima except with the express permission of the Railway and Canal Commission. The effect on the commercial flexibility of the railways was disastrous; increases in rates were not permitted and reductions were inadvisable because, if ineffective in stimulating traffic, they could not be retracted. This marks the complete ascendancy of trading interests with the railways being regulated as a public service to them.

THE DECLINE OF THE CANALS

In their dealings with the canal companies the railways were themselves accused of exploiting an advantageous position. Once the railways had proved their technical superiority the canal companies resorted to the strategy of objections to railway promotions. Faced with the inordinate expense of an opposed bill it was often cheaper for railway promoters to buy out the objecting canal company than to fight it. As decline appeared inevitable canal proprietors were happy to sell, especially on the handsome terms available due to their high nuisance value.

It has subsequently been alleged that railway companies bought vital links in the canal system simply to destroy them and hence disrupt through traffic. In fact the Railway and Canal Traffic Acts of 1854 and 1873 ensured that this species of unfair competition was not practised. The Royal Commission on Canals and Inland Navigation of 1909 found that railway-owned canals were on the whole very well maintained but suggested that railway-owned canal tolls were sometimes designed to divert traffic to the railways rather than attract it to the canals. Though the decline of canal transport may thus owe something to unfair competition, other factors, such as the technical superiority of the railways and the failure of canal managements to produce an adequate network by standardization of facilities and technical improvement, were of much greater significance as explanations of the decline.

THE REVIVAL OF ROAD COMPETITION

The second half of the nineteenth century saw signs of the revival of road transport, the course of which was severely curtailed by restrictions imposed as a result of public policy. Horse-drawn coaches proved inferior to the railways for the carriage of passengers on the grounds of speed, capacity and eventually comfort. Steam driven coaches might have proved to be the answer to some of these deficiencies but they were bitterly opposed from the outset. The Turnpike Trusts effectively outlawed them by charging tolls ten or twelve times as high as those for horse coaches. The result was that the steam coaches, first put on the road in the early 'thirties, were nearly all withdrawn by 1835 except for the haulage of very heavy loads. By 1865 the maximum permissable speed had been reduced to 4 m.p.h. and each vehicle had to be preceded by a man carrying a red flag. In addition to this crippling restriction the Highways and Locomotives Act of 1878 required a licence fee of £10 to be paid to each county council within whose area the vehicle was to operate.

The advent of the internal combustion engine led to great discontent with these restrictions which were finally abolished for vehicles of less than 3 tons weight by the Locomotives on Highways Act, 1896. Even the speed limit was raised to 14 m.p.h. The weight limit continued to restrict the provision of public road transport services until, in 1905, a departmental order issued by the Local Government Board under the provisions of the 1903 Motor Car Act raised both the speed limit and the maximum weights to which the limits applied. Road transport was immediately transformed by the rapid replacement of horses by motor vehicles.[1]

The early objections to steam traction on roads were not without reason. The road improvements of Telford and Macadam had, by 1820, made quicker travelling possible on the roads, but were not sufficient to stand the greater weight involved in steam traction. The Turnpike Trusts, which provided the majority of trunk road facilities, were commercial entities aiming to make profits; steam traction promised to increase costs but not significantly to increase revenue. Hence the higher tolls which drove the steam engines off the roads.[2]

Though their attitude to steam traction was understandable, the Turnpike Trusts cannot be entirely absolved from blame for the retarded development of road transport. Many of them were notoriously inefficient in providing even for horse-drawn transport. In some cases no funds were left for road maintenance after the payment of interest and administration expenses so that the responsibility fell back on to the parishes. The replacement of the statute labour system by a system of highway rates in 1835 transformed this into a financial strain on the parish rate. The decision to allow the Turnpike Acts to expire as rapidly as possible after 1864 in-

[1] By October 1911 the London General Omnibus Company had completely replaced its 17,800 horses by motor buses.
[2] See E. A. Pratt, *History of Inland Transport*, p. 485.

creased this burden to such an extent that the central Government first made a grant to the parishes to cover part of the expense and finally, by the Local Government Act of 1888, transferred the responsibility for all main roads to the county councils. The rapid development of mechanically propelled road vehicles after 1896 accentuated the deficiencies of the road system. To meet the need the Development and Road Improvement Funds Act, 1909, constituted a body known as the Road Board with powers to make grants to highway authorities for new construction or improvements, or to construct roads on its own account. The Road Improvement Grants were financed by the imposition of motor spirit duties and licence fees. Thus by the outbreak of the First World War the Government had accepted responsibility for ensuring that the development of vehicles and of roads for the vehicles to use were kept suitably in phase.

Tramways

In one other form of transport, the tramways, we can discern on a local scale the same motives for restriction which applied to the railways on a national scale. Local gas and water companies had established monopolies which had offended the local authorities. Consequently when tramways came into vogue Parliament was persuaded to pass the Tramways Act, 1870, which was designed to prevent any repetition of this monopoly exploitation. A tramway company had to obtain permission to construct both from frontagers and from the local authorities; moreover any concession was only valid for twenty-one years, at the end of which the local authority had the option of acquiring the concern at the physical value of its assets, with no regard to goodwill and profitability. This act seriously retarded the introduction of electric tramways, for companies were unwilling to undertake the heavy capital investment required in the knowledge that they could be taken over at the end of twenty-one years. Furthermore, many local authorities would only give permission for the construction of tramways on payment of large sums for street improvements, often quite apart from the proposed routes of the tramway. This discouragement of private investment became so severe as to necessitate municipal enterprise if tramways were to be widely provided. Unfortunately, no sooner had many municipalities constructed tramways than they were outdated by the invention of the trolleybus and the motor omnibus.

THE FIRST WORLD WAR AND TRANSPORT POLICY

The establishment of the Road Board in 1910 had been a major step towards the development of a national roads policy and, despite the relatively meagre resources at its disposal, by 1914 it had considerable achievements to record. This programme had to be curtailed during the war. The development of mechanically propelled vehicles was, on the contrary, accelerated by the war with the result that the respective development of vehicles and roads became seriously out of phase. At the end of the war, therefore, the authorities were faced with the problem not only of making up arrears but of getting the road system up to a higher standard of construction and repair than had previously been necessary.

A similar deterioration of physical capital overtook the railways. Though not subject to air attack as in the Second World War, their resources were nevertheless seriously denuded. During the war they lost about 30 per cent of their staff; material, locomotives and rolling stock were sent in large quantities to the western front and the railway workshops were transferred from ordinary maintenance to special war work. Hence the whole transport system was destined to be faced with a serious problem of re-equipment in the post war period.

One beneficial effect of the war was to unify the railway system. At the outbreak of war the Government assumed control of the railways which were, however, operated by a Railway Executive Committee composed entirely of managers of the existing companies. The operating problems had clearly been given adequate prior consideration for, despite the increasing strain imposed on decreasing manpower and resources, the railways achieved all that was required of them. The financial aspects were less satisfactory. All naval and military traffic was carried free from the outset and during the war this was extended first to the Ministry of Munitions and then to all Government departments. In return the Government guaranteed to the railways their 1913 net revenue. Apart from a 50 per cent rise in fares to discourage passenger travel charges were not increased. Wage levels did rise however and by the end of the war the wage bill was nearly three times its prewar level. The result was that immediately the railways were returned to peacetime operation instead of being able to make up the arrears of maintenance and investment they were faced with alarming current deficits. Unified operation of the railways during the war had clearly yielded great technical benefits yet they were financially in a far worse position than ever. The State, which had used the railways to such good purpose during the war, could hardly escape the responsibility to see them return to a sound footing.

THE INTER-WAR PERIOD

The first sign that transport policy was to be a matter of serious concern to the Government came with the creation of a separate Ministry of Transport in 1919.

The first major problem to be tackled was the restoration of railway finances. In an attempt to achieve this, compensation was paid for any deterioration of assets during the war period and 100 per cent increases in rates were authorized. These first steps proved to be totally insufficient, for the coal strike of 1921 hit rail traffic so badly that a total deficit of £60 million was incurred in that year alone. It was in these straightened circumstances that the railways were returned to private control by the 1921 Railways Act.

The Railways Act, 1921
The Act falls in three parts:

(a) The amalgamation provisions, whereby railway assets were vested in four regional companies.
(b) The provisions for the centralized determination of wages, hours of work and conditions of service by the newly created Central and National Wages Boards.
(c) The regulations for railway rates and charges. A new classification and schedule of standard rates was to be determined. This was not henceforth to be significantly altered without the consent of the Railway Rates Tribunal except in the case of reductions of up to 40 per cent of the standard, these rates to be known as 'exceptional rates'.

It was immediately pointed out that the separation of control of wages from the control of charges could well produce incompatible approaches which would seriously strain the finances of the companies.[1] But this was not the main source of strain in the depressed trading conditions of the 'thirties. The newly determined rates structure was almost totally ineffective from the outset. The number of 'exceptional rates' grew and by 1935 they accounted for 84 per cent of the tonnage carried. Even the standard rates did not prove to be as rigid as anticipated for the Tribunal had been given the overall duty of authorizing rates which would enable the railways, with

[1] See W. M. Acworth, *Elements of Railway Economics*. New revised edition, Oxford, 1924.

reasonable operating efficiency, to secure a 'standard revenue' defined as the net revenue in 1913 plus additions for extra capital or efficiency acquired since that date. In face of road competition and declining trade the railways never secured this revenue and the Tribunal proved very susceptible to any rates proposal aimed at achieving the standard revenue. Thus the standard rates could be altered with comparative ease.

Though this increased the flexibility of rail charges to a certain extent there were other statutory requirements which made railway charging very inefficient in face of road competition. Commercial freedom was limited by the liability to disaggregate all rates on demand, by the law restricting undue preference which inhibited the quotation of low rates to obtain traffic lest all similar rates had to be reduced accordingly, and by the classification system under which all valuable traffics were highly rated irrespective of the cost of transport. Largely as a result of these restrictions the railways lost over one third of their general merchandise traffic in the inter-war years.[1]

The control of road transport

The new Ministry of Transport was also responsible for organizing an effective national road system. A Special Fund of £10½ million had been set aside in 1918 to assist highway authorities to undertake maintenance work postponed during the war. The Roads Act, 1920, supplemented this by setting up a standing fund known as the Road Fund into which all licence fees and duties on petrol were to be paid and from which the roads department of the Ministry was to be supported and grants made towards the maintenance of roads by county councils. This structure of highway administration was considerably altered by the Local Government Act, 1929, but the system of grants continued. County councils received proportionate grants towards the approved cost of Class I and Class II roads whilst the large boroughs received provision for roads as part of a block grant.

So much for the immediate problem of the roads themselves. The rapid increase in traffic on these roads during the 'twenties gave rise to concern both because of the fear of danger to public safety by heavy and relatively uncontrolled road traffic and because it was suspected that the expansion at the expense of the railways was being artificially fostered by dissimilarities in the forms of control of the two sectors. Consequently in 1930 a Royal Commission on Transport was set up to consider three major problems:

[1] The tonnage of general merchandise carried by the railways declined from 68·4 million tons in 1919 to 44·3 million tons in 1938.

(i) Road traffic control and safety.
(ii) The control of public service vehicles and the regulation of road passenger transport.
(iii) General co-ordination and the development of all available means of transport.

Separate reports were issued on each set of questions. The recommendations of the first two reports were embodied, with commendable speed, in the Road Traffic Act of 1930, the provisions of which are discussed in detail in later chapters. The third and final report, dealing with the general problem of co-ordination and development, was a notable failure. Its suggestions, which included some redistribution of the burden of road taxation and a licensing system for road haulage subject only to fair wage and safety requirements, lacked real conviction. Moreover, when it came to the main problem of co-ordination it explicitly admitted that it had no useful contribution to make. The Government therefore turned to the industry itself and set up a committee of road and rail representatives under the independent chairmanship of Sir Arthur Salter with instructions to suggest a fair basis of competition between road and rail. The recommendations of this committee, embodied in the Road and Rail Traffic Act, 1933, form the basis of the current road haulage licensing system.

The consequences of these acts were not entirely satisfactory. The restrictions imposed on entry to road haulage reduced the number of bankruptcies in this field and made it a relatively comfortable industry for existing operators. Road safety was also improved, though this was probably due more to the safety requirements of the 1930 Act than to the licensing provisions of 1933. For the railways, however, decline continued, accentuated by the depression in the heavy industries. Even the most profitable of the companies, the Southern, only paid a dividend of $\frac{1}{2}$ per cent in 1938. The L.N.E.R. paid no ordinary dividend at all between 1925 and 1938.

The railways still attributed this situation to the form of control to which they were subject. Discontent concerning the apparent disparity between the statutory restrictions on the railways and the comparative freedom of road operators thus reappeared in the 'Square Deal' campaign of 1938. The railway companies sent a memorandum to the Government asking for repeal of the regulations of railway charges and also for freedom to make their own classification and conditions of carriage. The Government passed the claim to the Transport Advisory Council which referred it back to the industry with the suggestion that a series of bilateral conferences be set up between the railways and their main competitors and customers to negotiate a mutually acceptable solution. Some uneasy compromises were agreed but before it was possible to consider any legislation to put them into force the outbreak of war saw the railways once more under Government control and the immediate need for such legislation had passed.

An entirely different approach to the problem of co-ordination in the inter-war years was embodied in the scheme for London Passenger Transport. The original plan put forward in 1929 as a joint proposal of the Underground companies and the L.C.C. was rejected by Parliament but a similar scheme was enacted four years later as the London Passenger Transport Act, 1933. This act established the London Passenger Transport Board with a local monopoly in passenger transport excluding only the taxis, the main line railways and contract coach work. The apparent success of this experiment in co-ordination by administration rather than through the market was taken by many to be a good precedent on which to base an approach to the national transport problem after the Second World War.

THE SECOND WORLD WAR

Government occupation and operation of the railways was very similar in the two world wars. A Railway Executive Committee, composed of railway managers, was responsible for operation of the system and a fixed annual net revenue was paid.[1] Similarly, though not until mid-1942, the Government took control of all canals deemed capable of significant contribution to the war effort.

Control of road transport took a different form. Though the objectives of control were relatively simple, namely to restrict road transport to short distances in order to conserve petrol supplies, the predominance of small operators in the industry made implementation of any policy difficult. Fuel rationing was used to restrict mileage but vehicles were so rarely concentrated in the desired locations, and it was so difficult to induce small operators to change their base temporarily, that the Ministry of Transport was ultimately forced to operate a large fleet of vehicles on charter as a mobile workforce. The rest of the industry was then organized into regional pools through which any residue of Government road traffic could be allocated.

The bulk of the war transport burden thus fell on the railways. Freight traffic increased by one-third and passenger traffic nearly

---

[1] A profit sharing incentive, tried out at the beginning of the war, was soon abandoned as unsatisfactory.

doubled whilst the stock of locomotives, wagons and coaches remained approximately the same. An increased proportion under or awaiting repair reduced the effective stock and by 1945 even the condition of the effective stock had deteriorated drastically.[1]

The period of wartime control and direction by the Government again demonstrated the technical advantages of a unified transport system which the Labour Government endeavoured to retain by the Transport Act, 1947. This act constitutes the first attempt to treat the transport problem as a single integrated problem. It represents a predisposition towards planning which distinguishes it from all previous instruments of public policy for transport and we shall consider it in a separate chapter.

CONCLUSIONS

We may discern several motivations for Government intervention in transport, not all of which are mutually compatible. For instance, we may cite:

(a) Concern for the physical safety both of employees and of the public.
(b) Control of monopoly, both to restrain prices and profits and to prevent inequalities of treatment between traders.
(c) Avoidance of the 'wastes of competition', particularly excess capacity and unnecessary duplication of services.
(d) Maintenance of standards of public service including the provision, under cross-subsidization, of certain unremunerative facilities.

Control, though extensive and detailed, has tended nevertheless to be hesitant, defensive and tardy. Edward Cleveland Stevens, writing at the beginning of this century on the relation between railways and the State, concluded: 'Throughout the history of railways Parliament has failed to devise a comprehensive scheme.'[2] Until 1947 this criticism had hardly been refuted.

[1] For a thorough description of the transport sector during the Second World War see C. I. Savage, *Inland Transport*. (History of the Second World War, United Kingdom Civil Series.)
[2] E. Cleveland Stevens, *English Railways*, London, 1915, p. 316.

CHAPTER SIX

Nationalization and Denationalization

THE 1947 TRANSPORT ACT

The Labour Party have, from their earliest days, been clearly committed to nationalization as the most likely solution to the transport problem. From their first thoughts on the subject, which concerned only the railways,[1] their aims broadened, with the apparent success of centralized control in two world wars and the experience gained in the nationalization of London transport, to include nationalization of all major public inland transport agencies.[2] The election, in 1945, of a Labour Government with a large majority in the House of Commons, made it possible to implement these plans. Despite the last ditch efforts within the industry itself to obviate the need for nationalization[3] the Transport Bill was introduced in November 1946, and received the Royal Assent in August 1947.

The new organization
The first part of the Transport Act, 1947, set up a body to be called the British Transport Commission with the general duty 'to provide an efficient, adequate, economical and properly integrated system of public inland transport and road facilities'. In order to carry out this duty the B.T.C. was given power to carry goods by road, rail and inland waterways, though it was also statutorily prohibited from operating taxis or expanding into such contiguous industries as motor repairing except as a purely ancillary service to their main operations.

The responsibility for operating the transport system was put into the hands of five separate executives covering railways, docks and inland waterways, road transport, London transport and hotels.

For the railways complete centralization of financial responsibility

---

[1] A. E. Davies, *The Case for Railway Nationalization*, London, 1912.

[2] H. S. Morrison: *Socialization and Transport*, London, 1933. H. S. Morrison, *British Transport at Britain's Service*, London, Labour Party, 1938.

[3] In July 1946 the Railway General Managers Conference and the Road Haulage Association together produced a memorandum entitled 'Co-ordination of Road and Rail Transport'. In this paper they suggested that co-ordination could be achieved under private enterprise on the basis of a published classification and rates structure for road haulage, similar to that operating for the railways.

coupled with a certain amount of regional flexibility in operating matters (which was to be exemplified by the use of different regional liveries) was decided upon. Such large scale organization was nothing new. Indeed, in a sense decentralization occurred, for there were now six operating regions instead of the four main line companies operating since 1921. Moreover, the complete financial amalgamation permitted these regions to be devised on a purely territorial basis without reference to either historical anomalies of ownership or prospects of separate financial viability.[1]

In contrast to the railways, the road haulage industry was totally unaccustomed to a highly centralized form of organization. It appeared to some critics that the technical advantages of large fleet operation might only be obtained at the expense of attention to the individual needs of customers and the spur of competition. Of course, a certain amount of competition would still be provided by those 'A' and 'B' licence carriers who were allowed to continue in private operation within a 25 mile radius of their operating base. In addition the Act contained two other provisions to secure the trader against arbitrary treatment or exploitation by a public transport monopoly, namely:

(a) By the operation of a consumer consultative machinery,
(b) By the continued freedom for any trader to obtain a 'C' licence to provide transport on his own account.

We shall discuss the effects of these measures later.

Acquisition and compensation
The provisions for transfer of assets to the B.T.C. were contained in Parts II and III of the Act. Railway securities were acquired at the stock exchange valuation prevailing in the week prior to the King's speech or at the average of the mid-monthly quotations for the six months preceding the general election of 1945, whichever was higher. This compensation was most generous as the stock exchange valuation at the end of the war reflected a considerable degree of optimism resulting from the large amount of traffic diverted to the railways during the war. Railway shares stood unusually high and by replacing the equity interest with fixed interest stock computed on this valuation the B.T.C. was committed to a far heavier annual payment to capital than had been earned by the railway companies during the 'thirties.

As so few of the road hauliers were large companies with quoted

---

[1] No separate Scottish company was formed in 1921 partly on the grounds that such a company would not have been financially viable.

shares a different method of compensation had to be used for them. Physical assets were acquired at replacement cost less a compounded 20 per cent per annum for depreciation and the goodwill of the business was paid for separately in a lump sum of not less than twice nor more than five times the net annual profit. Nationalization deprived owner-managers not only of income but also of employment and though many of them were re-employed by the Road Haulage Executive a loss of status was frequently involved.

The financial and commercial obligations of the B.T.C.
Financial responsibility and control was centralized in the hands of the B.T.C. itself rather than committed to the separate executives. Further to the general duty of the B.T.C. to provide an adequate, efficient and properly integrated system the Act added the financial requirement that 'the Commission shall so conduct its undertaking as to secure that revenue is not less than sufficient for meeting costs properly chargeable to revenue, taking one year with another'. This leaves a considerable vagueness concerning exactly what are the costs 'properly chargeable to revenue' and the period over which the budget must be balanced. However, the general duty to provide an 'adequate' service did indicate the continuance of many unremunerative services. Indeed, the fact that the executives were not separately accountable made cross-subsidization possible on a very wide scale. On the other hand the requirement that the transport system should be 'economical and properly integrated' suggests that the B.T.C. was expected to devise some method of ensuring that each traffic was transported by the agency able to do the job most cheaply in real terms. But the Act also required that traders be able to choose freely between the alternative forms of transport. Consequently, as the B.T.C. could not direct traffic to one agency rather than another, it would have to secure the desired allocation of traffic by the inducement of a high standard of service or a relatively low price. But high standards of service can usually be provided only at high cost. Therefore, to secure an economic allocation of traffic by inducements of price or quality of service would seem to imply that the B.T.C. be left completely free to relate price to cost. In fact, both on the question of standard of service and on that of price the Commission was restricted.

The first restriction consisted of a machinery for consumer consultation, comprised of a central committee and area committees covering the whole country. The statutory function of these committees was to consider, and where necessary to make recommendations upon, any matters (including charges) relevant to B.T.C.

services and facilities. The provision of this machinery was in recognition of the fact that even a publicly-owned monopoly could act in an arbitrary manner, contrary to the public interest. In particular it was intended to protect the interest of small unorganized consumers against such arbitrary action. The machinery was left without teeth however. On the matter of charges the powers of the Transport Tribunal and the Minister seemed to render it redundant. Even in the matter of standard of service it was given no direct power. Recommendations from the machinery went to the B.T.C. and, in the case of the central committee, to the Minister. If the B.T.C. did not accept the recommendation only a ministerial directive could enforce it. Subsequent experience has shown that, whilst in minor matters the B.T.C. does respond to the promptings of the consultative machinery, it does not provide the means of enforcing any major change of policy on the B.T.C.[1]

The provisions concerning charges appeared, on the surface, no more likely to restrict railway management seriously. There was no detailed specification of how charges were to be determined and freedom was limited by only three restraints. Firstly, there was the general financial requirement that, in total, revenue should cover costs, taking one year with another. Secondly, was the ministerial power to issue directions of a general nature, which could include directions on rates and charges. Finally, in Part V of the Act, it was provided for the supervision of rates and charges by a Transport Tribunal. The B.T.C. was obliged to prepare comprehensive charges schemes and submit them to the Tribunal. Both consumers and competitors had rights of objection, and, after hearing these objections at a public inquiry, the Tribunal could accept, reject or alter the schemes as it saw fit. The Tribunal was itself subject to certain limitations. Section 82 of the Act enabled the Minister to authorize provisional increases of charges while a decision was pending. Moreover, Section 85 required that neither the Tribunal nor the B.T.C. should take any action which would prevent the B.T.C. from making its proper revenue, taking one year with another. These provisions appeared to protect the B.T.C. against any intervention likely to make it incur losses. In practice, however, the phrase 'taking one year with another' has been interpreted very broadly by both the Tribunal and the Minister to justify postponing, modifying or even rejecting charges proposals despite current deficits.

[1] See G. Mills and M. Howe, 'Consumer Consultation and the Withdrawal of Railway Services', *Journal of Public Administration*, Vol. 38, No. 3, Autumn 1960; also K. M. Gwilliam, 'Consumer Consultation in the Nationalized Industries', *Applied Economic Papers*, Vol. 2, No. 2, September 1962.

G

Despite these restrictions the 1947 Act was still in principle superior to the 1921 Railways Act in that it gave the initiative in preparing the form of the charges schemes to the B.T.C. without any hindrance. Unfortunately this did not result in the introduction of charging methods satisfactorily designed to secure an economic allocation of traffic for two reasons.

In the first place, the financial provisions of the Act were sufficiently flexible to permit the continuation of the various forms of cross-subsidization which had become firmly entrenched as railway pricing principles. Though charges discriminating in favour of bulky, cheap goods and against light, expensive commodities were not related to cost the Transport Tribunal was unlikely to yield easily to an attempt to radically revise such a generally accepted system. Heavy and recurring losses might have broken down the resistance of the Tribunal on this question, but the grant of an almost total monopoly of public transport to the B.T.C. seemed to ensure them against such an eventuality. Despite the monopoly element losses eventually occurred due to the extension of 'C' licence operation and, after 1953, due to competition from the denationalized road haulage industry. Cross-subsidization could only be successfully practised as long as the profitable as well as the unprofitable traffics could be secured by the B.T.C. Even before denationalization of road haulage many of the traffics most profitable to the railways could be carried more cheaply by 'C' licensed road haulage. Hence much of this traffic was lost and the bulwarks of cross subsidization disappeared.[1] At the same time, however, even those traders operating 'C' licensed vehicles would find it profitable to send their more difficult consignments by rail to take advantage of charges below cost for certain services. The resulting allocation of traffic did not necessarily correspond with that which would have ensued if all rates had been more sensibly related to costs.

Secondly, the preparation of a draft charges scheme for road haulage was a problem that the B.T.C. did not succeed in solving. Whereas a railway rates structure was already in existence as a starting point, there had never been a national road rates structure, despite discussions along these lines in 1938 in the course of the 'Square Deal' negotiations and in 1945–6 prior to nationalization.

[1] In the original Transport Bill it was proposed to confine the 'C' licence operator to the same restricted radius of operation as the 'A' and 'B' licence holder. Some Labour M.P.s argued that even this restriction, which was subsequently dropped, was insufficient to prevent the abstraction of the more profitable traffic from public transport. See, for instance, the remarks of Mr Ernest Davies during the Second Reading debate—*Hansard House of Commons Debates*, 1946–7, Vol. 431, Col. 1,652.

In any case it is doubtful whether a national schedule would have fitted the bill at all. Costs for road haulage are governed by specific conditions such as loading characteristics, route characteristics, availability of return loads, etc. which vary from traffic to traffic and from time to time but which can nevertheless be identified with considerable degree of accuracy before the traffic is accepted. An enforceable national rates structure would have necessarily eliminated many of these subtleties, thus further divorcing the price of transport services from the cost of provision.[1] By 1952 when denationalization of road haulage was announced, no national rates structure had been introduced. During the preceding four years road charges had been based largely on pre-nationalization schedules relating to the costs incurred while rail charges were based on a far less accurately cost orientated procedure. Clearly, so long as these dissimilar pricing techniques were used the proper integration of the alternative forms of transport was not compatible with the complete freedom of choice of agency permitted to all customers.

The provisions for road passenger transport
The main line railway companies, when faced with road competition in the inter-war period, had met it in many cases by buying bus companies. Thus, when the assets of the railways were transferred to the B.T.C., quite a large road passenger transport system came under Commission control as well. In addition to this the B.T.C. used the powers granted to them under Section 2 (2) f. of the Act to acquire the Thomas Tilling group, the road passenger transport undertakings of the Scottish Motor Traction group, and certain smaller undertakings.[2] After the completion of these purchases the two consolidated groups were retained as units of management of the road passenger transport interests until such time as the area schemes, provided for in Part IV of the Act, could be brought into operation. In the meantime these B.T.C. owned undertakings were partially exempt from the normal licensing requirements, though they were only to operate on routes approved by the Traffic Commissioners.

Part IV of the Act empowered the Commission to prepare area passenger road transport schemes for submission to the Minister. These schemes would specify what services were to be provided in any area and who was to provide them, arrange for the transfer to public ownership of specified undertakings and regulate the relations between road and rail passenger transport, including, where necessary,

---

[1] See R. Cropper, 'Road Freight Transport and the Act of 1947', *The Manchester School*, Vol 18, No. 3, September 1950.

[2] Principally the British assets of the United Transport Group of Chepstow.

the pooling of receipts and expenses. If such a scheme damaged the interests of an existing operator then either his undertaking was to be acquired by the B.T.C. or just compensation was to be paid for his loss. The actual wording of the Act does not make it clear whether the schemes envisaged were to involve the acquisition of all road passenger transport undertakings or not. Opposition critics contended that, as only permissive powers were granted in respect of road passenger transport, the Government was itself undecided as to the desirability of nationalizing this sector. It seems to have been intended, however, that eventually the B.T.C. might provide all types of road passenger services, including contract services, though it was possible that some small operators might have been left outside the nationalized sector. The vagueness of the Act in this respect probably represented room for manoeuvre in face of any difficulties which might arise rather than any vacillation in principle.

The Road Transport Executive began by provisionally dividing Great Britain into suitable areas for the schemes and as a first essay proceeded with the preparation of a scheme for the Northern Area, comprising Northumberland, Durham and much of the North Riding.[1] As the extension of B.T.C. interests proceeded in 1949 it was decided that a separate Road Passenger Executive be set up to develop the area schemes. Hence, from June 20, 1949, the responsibilities for road haulage and road passenger transport were separated. By the end of 1951 the Road Passenger Executive was able to report that it had submitted the scheme for the Northern area to the Minister and that it also had in hand schemes for East Anglia and the southwest. Before any of these schemes could be completed the new Conservative Government announced, in April 1952, its intention to pass a new Transport Act which would repeal the provisions relating to road transport. A Bill duly followed in July with the result that work on the schemes was discontinued and the executive disbanded in October.

THE 1953 ACT

Many of the provisions of the 1947 Act were hardly in operation before they were negated by the contents of the Transport Act, 1953. There are two sets of provisions in this latter Act, those relating to road transport and those to rail; they are very closely connected in that the rail provisions were a *quid pro quo* for the denationalization of road haulage.

[1] The progress of this scheme can be traced in the B.T.C. Annual Reports which mentioned it each year (1949, p. 14, 1950, p. 15, 1951, p. 7).

In the first draft of the Transport Bill it was proposed that the assets of the Road Haulage Executive should be sold up piecemeal and that any loss on the sale, together with the loss incurred by the B.T.C. consequent on the return of road haulage to private ownership, should be met by a levy imposed on all road transport operators. 'C' licence operators immediately objected to paying part of the cost of returning road haulage to private enterprise. The Bill was therefore modified so that the levy on road haulage was only to cover the loss on resale and not any subsequent operating losses of the B.T.C. Furthermore the B.T.C. was prevented from purchasing any more road passenger services except by express consent of the minister, who was further empowered to require the Commission to dispose of their existing holdings in bus and coach companies if he saw fit. Advantage has not so far been taken of this permissive power. At the same time control of road transport was returned to the Licensing Authorities and Traffic Commissioners acting under the provisions of the legislation of the early 'thirties.

In order to counterbalance any deterioration in the railway's financial position resulting from the denationalization of road haulage it was enacted that they should be relieved of some of their traditional restrictions. They were no longer required to avoid undue preference or to show equality of treatment between traders. The Transport Tribunal was to have control over a framework of published maximum charges but within these maxima the B.T.C. was free to determine its charges as it wished, without publication. There were some notable exceptions to this freedom. Where railways still exercised a virtual monopoly of transport of a commodity the Tribunal retained the power, on appeal by any trader, to determine a 'reasonable' charge. Road hauliers were also granted rights of appeal to the Tribunal where it could be shown that particular charges would result in loss to the Commission and had only been quoted in order to eliminate competition.

THE CONSEQUENCES OF NATIONALIZATION

The 1947 Act set its own criteria by requiring the B.T.C. to provide an efficient, adequate, economical and properly integrated transport system. As far as efficiency and economy are concerned judgment is very difficult not only because of the conceptual problems involved but also because the structure was changed too soon for the full consequences of the system to have materialized. The railways were still starved of capital due to the decision of the Government to concentrate early post war investment in other sectors. Similarly in road transport, four years was totally insufficient to make a balanced judgment on the merits of the organization. When the Road Transport Executive came into being in 1948 it had neither vehicles nor personnel. During the first year the 400 largest concerns were voluntarily transferred to form the nucleus of British Road Services. By the end of 1951 compulsory acquisitions had enlarged this fleet to one of about 41,000 vehicles and B.R.S. could boast a virtually complete network of trunk services. This alone was no mean achievement. Even the problems of organization and management were being handled in an encouraging manner, with many former private hauliers settling down as efficient and enthusiastic group managers. The charges of inefficiency of operation made against the Road Haulage Executive to support denationalization were thus largely unfounded.

A far more justifiable criticism of the organization of transport under the 1947 Act would have been that it did not contain anything more than pious hope as far as the 'proper integration' of transport was concerned. The Minister of Transport had argued in the House in 1947 that one of the most beneficial results of the Bill would be to 'simplify the jungle of rate charges that now prevails in our railway and transport industry'.[1] But a proper integration of transport required not that the rate structure be necessarily made simple, but that it be made rational in relating charges in some sensible way to costs. The failure to produce a new charges scheme which would meet this need was probably the most serious and fundamental deficiency of the nationalized transport sector.

The adequacy of transport services is another difficult quality to adjudge. Clearly the provisions for cross-subsidization in the 1947 Act implied a political judgment concerning the desirability of maintaining certain unremunerative services. This partly represents an evaluation of social costs and partly a comment on the distribution of income. These are value judgments which are susceptible to approbation or disapprobation but not to strict economic proof or disproof. We might still argue however that the 1947 Act was likely to produce an inadequate service on other grounds. Transport is a very heterogeneous service. The larger the number of alternatives the greater is the chance of any individual finding one tailored to his particular needs and tastes. Now admittedly too many small operators in competition may provide a framework which is unstable and therefore in the long run fails to maintain the service desired by consumers, but on the other hand the restriction to two alternatives, the nationalized road and rail carriers, may be unnecessarily severe.

[1] *Hansard House of Commons Debates*, Vol. 431, Col. 1,633.

The Act of 1953 was based on the dogmatic premise that this was an undue restriction for the Conservatives were less enamoured than their predecessors with the technical advantages of a large trunk network of road services. But if this was the case denationalization was not the best way to demonstrate it. As Professer Walker has argued, 'Large scale in itself is neither good nor bad. If one large publicly-owned undertaking can provide as economically and efficiently as a number of small concerns then there is no good economic reason for preferring one to the other. The menace of nationalization in road transport lies not in the size of organization nor in the question of ownership, but in the element of exclusive monopoly, enjoyed by B.R.S., of public long distance carriage by road. That monopoly could readily have been destroyed by the simple means of lifting the 25 mile limit and, had the Conservatives really had the courage of their convictions, by removing the requirement of an "A", "B" or "C" licence as imposed by the Act of 1933. Road Transport would thus have been opened to private competition and the commercial efficiency of the small private undertaking would have been restored without losing whatever may be the technical efficiency of the larger.'[1]

As it happened unexpected difficulty was encountered in selling off the B.R.S. assets. As a consequence of this the Transport (Disposal of Road Haulage Property) Act 1956 permitted the B.T.C. to retain about 7,000 lorries to maintain the trunk network. The fact the B.R.S. has continued to operate successfully in competition despite the loss of much of its best equipment and man-power at the time of denationalization, suggests that the technical advantages mentioned by Professor Walker were quite significant.

The return of road transport to private ownership meant that there could be no cross-subsidization between road and rail. Thus, though in a sense it is no more important now than it always has been as far as the proper co-ordination of transport is concerned, it becomes more obviously crucial to the financial viability of the separate undertaking to ensure that the charges of transport operators be related to the real cost of the service provided. The third part of the book, by considering the current organization and principles of operation of the industry, will help us to assess the extent to which this is being achieved.

[1] G. Walker, 'Transport Policy Before and After 1953', *Oxford Economic Papers*, Vol. 5, No. 1, March 1953.

PART III

THE AGENCIES

CHAPTER SEVEN

Road Passenger Transport

It has been estimated that roughly 80 per cent of all passenger miles travelled in this country are, in one form or another, travelled by road.[1] Within this sector there are several different types of service.

The most obvious distinction is that between private transport (car, motor-cycle, etc.) and public transport (bus, trolleybus, tram). Though accurate figures are not obtainable it was estimated in 1958 that 46 per cent of road passenger transport was provided by private means. Moreover, this proportion is increasing. In Britain as in all the countries of Western Europe, the number of cars has been increasing rapidly over the last decade. The limit does not appear to have been reached yet for whereas in the U.S. there is one car for every three people the ratio in Great Britain was still, in 1960, only one car to every 9·2 people. The result of this trend is an increasing number of vehicles per road mile but a decreasing proportion of these vehicles being engaged in public transport of any kind. Indeed in Britain not only the proportion but the absolute number of vehicles providing public facilities is declining (see Table 3).

TABLE 3

Vehicles with licences current in Great Britain

	1938	1948	1958	1961
		'000	*vehicles*	
Private cars	1,944	1,961	4,549	5,979
Motor-cycles	462	559	1,520	1,869
Total private transport (1)	2,406	2,520	6,069	7,848
Bus and coach	50	65	75	76
Trolleybus and tram	12	10	4	2
Taxi	35	59	17	13
Total public transport (2)	97	134	96	92
All passenger vehicles (3)	2,503	2,654	6,165	7,940
(2) as per cent of (3)	3·9	5·0	1·6	1·2

We will begin therefore with a brief consideration of the private sector.

[1] Select Committee on Nationalized Industries, 1960. Appendix 8, p. 335. According to this estimate 25,000 million of a total of 120,000 million passenger miles were travelled by rail in 1958.

PRIVATE MOTORING

The costs of private motoring may be divided into the normal economic categories of fixed costs and variable costs. The items falling into each of these categories are shown in Table 4.

TABLE 4

The Cost of Motoring

Fixed Costs (£ per annum)	I		II		III	
	£	s	£	s	£	s
(i) Interest on capital	30	0	75	0	5	0
(ii) Depreciation	90	0	225	0	15	0
(iii) Motor and driving licences	15	5	15	5	15	5
(iv) Insurance	25	13	42	12	8	0
(v) Garaging	45	10	45	10	–	–
(vi) Subscriptions	2	2	2	2	2	2
Total	208	10	405	9	45	7
Running Costs (pence per mile)						
(vii) Petrol	1·49		2·93		2·20	
(viii) Oil	·13		·21		·27	
(ix) Tyres	·21		·32		·21	
(x) Servicing	·41		·52		·41	
(xi) Repairs and replacement	—		—		1·50	
Marginal cost (pence per mile)	2·24		3·98		4·59	
Average Cost (pence per mile)						
Running 1,000 miles per annum	52·28		101·29		15·47	
Running 5,000 miles per annum	12·25		23·44		6·77	
Running 10,000 miles per annum	7·24		13·71		5·66	
Running 15,000 miles per annum	5·58		10·47		5·32	

Of the specimen costings shown in this table, column I represents the costs of a new small family car costing £600 and running 40 m.p.g., column II those of a new large car costing £1,500 running 20 m.p.g., and column III those of an old small car costing £100 and running 25 m.p.g.

Some of the cost heads require explanation. Firstly, the capital costs. It is only sensible to impute as a cost of motoring some interest on the capital involved. In our examples we have used 5 per cent as the relevant rate, though the rate paid on the hire purchase of a car would be considerably higher than this. Similarly with garaging, though most motorists provide their own garage accommodation, for cars I and II we impute the value of garaging at the market price which the A.A. estimates as 17s 6d per week.[1]

[1] Several of the above estimates are based on the A.A. Schedule of Estimated Vehicle Running Costs (September 1961).

Depreciation may be measured in terms of the decline of the market value of the vehicle during the relevant period, though if the general price level of cars is rising we ought to make extra provision to ensure replacement. It is included as an overhead cost because, though partly a function of vehicle mileage, it is most significantly a function of age of vehicle. We assume in our examples depreciation of 15 per cent per annum on the market value of the vehicle at the beginning of the year.

For the new cars (I and II) we assume that comprehensive insurance cover is purchased whereas for the older car third party cover only is taken.

The variable costs are so called because they are directly related to vehicle mileage, even though the incidence of some of the costs may be irregular (e.g. servicing every 1,000 miles, tyre life 20,000 miles). Fuel, oil and repair costs are also variable, however, with respect to the age of the vehicle; in our examples we have assumed repair costs to be nil for the new cars and to be 1½d a mile for the older vehicle.[1]

The estimates demonstrate clearly that, because of the incidence of capital costs and depreciation, the average cost per mile run by the newer cars is significantly greater than that of the older car for mileages less than 10,000 per annum.[2] For mileages above 10,000 per annum the difference in cost per mile is unlikely to be sufficient to compensate for the differences in comfort and reliability.

The decision to purchase a car, if considered carefully at all, is likely to be made on grounds of the expected total cost, limited perhaps by the availability of capital. Subsequent decisions concerning use of the car are more likely to be made, however, on the basis of marginal costs. Indeed they may be made on the basis of fuel costs only as it is easy to forget those variable costs such as tyre wear and servicing costs which do not arise as immediate out-of-pocket expenses. The marginal cost per vehicle mile does not vary greatly according to the number of passengers carried. Therefore the cost per passenger mile can be regarded roughly as the cost per vehicle mile divided by the number of passengers. Thus for car I in our example the average variable cost per passenger mile for a married couple travelling together at less than 1¼d is below the ordinary rate either by rail or by bus.

But cost is not the only determinant of car ownership and use. Car travel is a far more convenient and flexible mode of transport

[1] This is compatible with the A.A. estimate that the average repair cost for a 1 litre car over its whole life is ¾d per mile.
[2] Such calculation would be far more complicated if differential road charges were applied as described in Chapter 12.

than any public form. On the other hand it may be slower than rail transport for some journeys and driving, though still a pleasure for some, can be a tiring and harrowing exercise in heavy traffic or on a long journey. On balance these considerations are generally assumed to favour motoring; this advantage will be diminished if congestion increases.

PUBLIC ROAD PASSENGER TRANSPORT

Though there are several different types of public road passenger transport services (stage, express, unscheduled excursion and tour, contract) with differing characteristics, there are certain major features of the costs of operation which are common to all types of operator. As an example the approximate cost structure experienced by the Tilling group of companies in 1960 is shown in Table 5.

TABLE 5

Cost Structure of the Tilling Group 1962

	£000	%	
Vehicle Operating			
Drivers and Conductors	20,937	49·3	
Fuel and Oil	5,273	12·4	
Tyres	822	1·9	
Servicing, etc.	2,244	5·3	
Administration	437	1·0	
		29,713	70·0
Maintenance			
Repairs	4,186	9·9	
Depreciation	2,451	5·8	
Administration	298	·7	
		6,935	16·3
Other traffic expenses	3,002	7·1	
Maintenance of buildings	561	1·3	
Vehicle licence duties	316	·7	
General	1,919	4·5	
Total		42,447	100·0

Source: B.T.C. Annual Report & Accounts, 1962, Vol. II, Table IV–4, p. 87.

Even if we assume that neither repairs nor depreciation are directly variable with vehicle mileage, we still have almost 70 per cent of total cost directly variable with use.[1] Direct labour operating costs themselves account for nearly half of the total. Moreover, as both the

[1] The cost structure of bus operation is discussed in more detail in L. D. Kitchin, *Bus Operation*, Chapter 14.

group and the constituent companies are large and provide considerable terminal services it is likely that the proportion of direct runnings costs will be a little lower, and that of overheads higher, for Tilling than with a smaller concern. Within the group, however, the separate company cost structures are by no means homogeneous. In particular, costs will depend on the category of the services provided; for instance stage services, because they generally require a conductor, usually have the highest proportion of labour costs.

These costs are incurred not in providing a particular quantity of service but of capacity. The cost per unit of service will thus also depend on the load factor achieved. As the units of capacity are indivisible below the size of an individual bus the load factor will depend on demand, which tends to be very uneven in three respects:

(i) Between different times of day.
(ii) Between different periods of the year.
(iii) Between different geographical locations.[1]

Given these characteristics we can see how costs per passenger mile vary between types of service. Excursions, because they only run if demand is sufficient, and contract services, because they are priced on the cost of vehicle use, tend to have much fuller loads than stage services so that the running costs are spread over a larger number of passengers. On the other hand the seasonal fluctuation in demand destines the excursion operator to a far higher proportion of idle time so that the standing charge per unit of service sold will tend to be relatively high for coaches used solely for this type of service. Stage services, in contrast, have the lowest ratio of standing charges per vehicle mile but, because of the frequent stops, which increase fuel costs, and the need to carry a conductor, have the highest running costs per vehicle mile. As the running load factor will also tend to be low we get quite significantly higher running cost per passenger mile.

Labour costs form a high proportion of the total. It is therefore particularly important that the industry makes economical use of this factor. One man bus operation may offer considerable economy in wage costs on stage services though for urban running the absence of a conductor might lead to time losses which would seriously affect scheduling and certainly increase traffic congestion. Similarly it is most important that the labour employed should be 'productive'. An investigation undertaken in 1952 in the Scottish Omnibuses group[2]

[1] See G. J. Ponsonby, 'The Problem of the Peak with Special Reference to Road Passenger Transport', *Economic Journal*, Vol. 68, March 1958.
[2] J. Amos, 'The Operation and Economics of Road Passenger Transport', *Journal of the Institute of Transport*, Vol. 25, No. 2, January 1952.

revealed that over 25 per cent of labour time was non-productive in one way or another whilst 16·6 per cent of driver's time was paid for at rates above the standard (overtime, rest day employment etc.). Clearly cost reductions would ensue if some of the idle time could be used to reduce the proportion of excess rate work. Split shift working and tight scheduling might achieve this.

The cost of fuel is second only to labour in importance to the bus operator. With fuel constituting nearly 15 per cent of total costs it is obviously desirable to use vehicles which make the least possible call on the fuel as well as the labour resources available. Single deck buses offer greater scope for weight reductions and hence economy of fuel consumption but here public policy impinges crucially on the commercial decision. In the interests of limiting road congestion maximum dimensions are prescribed for public service vehicles. It is very difficult to increase the seating capacity of a single deck bus of the maximum body size (36 feet by 8 feet 2½ inches) to much more than 50, which may be considerably less than the desired vehicle capacity in peak periods.[1] Hence the double deck bus is likely to be retained for urban services because of its greater capacity and despite the fact that such catering for peak period traffic increases the costs per passenger mile at other times of the day.

Finally we may assess the importance of taxation as a cost of operation. It has been estimated that fuel tax accounts for between 9 and 12 per cent of the costs of bus operation.[2] To this we must add 1 to 2 per cent for licence duties to obtain an estimate of 10 to 14 per cent of total costs arising in this form.

The most important statistic of all is the cost per passenger mile. This will depend not only on the cost per vehicle mile but also on the load factor achieved. Although the operator may be able to entice custom to a certain extent by efficient operation, overall control of the relationship between demand and the amount of capacity provided is the specific function of the licensing system.[3] It is therefore

---

[1] Though Barton Transport operate some 56 and 57 seat single deck vehicles and the Ministry of Transport is currently examining the case for high capacity standee vehicles on the continental pattern.

[2] See the Report of the Committee on Rural Bus Services, *Ministry of Transport 1961*, Paragraph 148, p. 41.

[3] Competitive provision of services is controlled rather more precisely than capacity. Vehicle or seat limitations are rare on stage services where duplication by the licensed operator is possible within approved limits of the scheduled time. Similarly on express services there is usually a duplication allowance and an operator who consistently loads to full permitted duplication has a strong case for a licensed increase. Thus it is usually within the power of existing operators to vary capacity on their own routes.

fundamental to any investigation of the economics of bus operation to understand the way in which this system operates.

The licensing system

Prior to 1930 vehicles plying for hire were in theory licensed by local authorities. In practice many authorities did not exercise their powers and even where licences were issued the definition of 'plying for hire' was so vague that many buses were run subject to no control at all.

This led to two serious consequences. Firstly there was a great deal of excess capacity in the industry. Many buses and coaches were used only for very short periods of the year. A great wastage of capital resources was involved, for even during busy periods many buses were still operating with light loads. The second danger stemmed directly from this. Where no alternative employment was available, and in the inter-war period this was frequently the case, drivers were tempted to stay on the road as long as they could hope to obtain sufficient revenue to meet their out-of-pocket expenses. For owner drivers this meant anything more than fuel, tyre and servicing costs. Some operators were not even wise enough to require this and operated at a loss on running costs. In such a fiercely competitive situation dangerous driving practices were common in the attempt to secure a share of the limited traffic available. Moreover, buses that earned little more than running costs were allowed to deteriorate in condition until they were not only uncomfortable but positively dangerous. These problems made intervention desirable, though not necessarily in the form that it eventually took.

The Road Traffic Act, 1930, made it obligatory to obtain several authorizations before any service could be operated.[1] Firstly, each vehicle had to possess a certificate of fitness declaring it to meet the statutory requirements concerning construction, dimensions, equipment and general condition. Annual inspections were introduced and the certificate could be withdrawn if any defects became apparent. Secondly, the operator had to obtain a public service vehicle licence in respect of each vehicle; such licences would only be granted to applicants considered fit persons to operate public services. Thirdly, drivers and conductors had to obtain licences granted on the basis of their character as well as their technical competence.

These three types of licence together constitute a fairly tight control

---

[1] A fuller description is to be found in J. A. B. Hibbs, *The Effect of the Road Traffic Act, 1930, on the Development of the Motor Bus Industry.* Unpublished M.Sc. thesis, London 1954. See also J. A. B. Hibbs, *Transport for Passengers,* Hobart Pamphlet No. 23, 1963.

H

over the safety of vehicles and their operators. Effectively they were a once-and-for-all declaration of standards and did not form a continuing barrier to entry. Though in 1931–2 only 88·2 per cent of applications for P.S.V. licences were granted, by 1933–4 this had risen to 99·4 per cent and has been virtually 100 per cent ever since.

The real teeth of the new licensing system were contained in the requirement of a road service licence for every service, except contract carriage, on which passengers were carried. This licence states the type of service (stage, express, excursion), the route, the operator, the fare to be charged and any other special conditions which the Traffic Commissioners choose to attach. The country was divided into thirteen traffic areas and where a service ran through more than one of these areas licences had to be obtained for each (a 'primary' for the area in which the operator is based and a 'backing' for any other area involved).

The Act did not specify very exactly the grounds on which the Commissioners were to exercise their powers. They were not to issue a licence if it appeared that the speed limit of 30 m.p.h. was likely to be exceeded but otherwise it merely stated that, at their discretion, they might consider the suitability of the route, adequacy of existing services, the desirability of the service in the public interest and the general needs of the area (including the provision of some unremunerative services and co-ordination with other forms of transport). In order to achieve these ends the Act suggested that conditions might be attached to licences concerning fares, timings and the points at which passengers might be taken up or set down.

The structure of the licensing system is relatively simple. There are now eleven areas (following the abolition of the Southern area in 1934 and the merging of the two Scottish areas during the war). Each area except the Metropolitan has three Commissioners, the chairman being a permanent appointment of the Ministry of Transport, the others being appointed from panels of local authority nominees. In London there is only one Commissioner but in practice the chairman is also dominant in the other areas. All applications for licences appear in 'Notices and Proceedings', a document published weekly or fortnightly, according to the business of the area. Objections lodged within a fortnight will result in a public hearing before the Commissioners, otherwise the decision is made formally and published in a subsequent edition of 'N and P'. The Commissioners need not give reasons for their decisions except where appeal is made to the Minister. In such an appeal case each party has a copy of the Commissioners' statement and an Inspector, appointed by the Minister, hears the arguments of appellant and respondant concerning the decision. New evidence is rarely admitted at this stage. The final decision is made by the Minister on the basis of, though not always completely in accordance with, the Inspector's report. Ministerial statements of the considerations involved in appeal decisions, though forming neither a comprehensive nor a rigid case law, have been accepted as guidance for subsequent decisions. Three principles have emerged.[1]

Firstly there is the principle of *priority*. In making their original decisions the Commissioners, subject to the provision of a reasonable standard of service, laid considerable emphasis on the length of time for which the applicant had previously been providing the service concerned. Subsequently it has become normal practice for existing operators to be given preference over newcomers.

The second principle is that of *protection*. The first aspect of this is protection applied between types of service. Stage services are protected from incursions into local traffic by express operators in a variety of ways. The express operator may be prohibited from carrying short distance passengers within the stage service area, the express service minimum fare may be higher than the stage service maximum, or the shorter distance fares may be set higher for the express than the stage operator. In a similar way the regular traffic of the express operator is protected against the excursion operator by the imposition of detailed restrictions concerning timing, routing, fares, number of vehicles to be used and the type of passenger permitted to be carried by excursion operators.

The general aim of such protection is to prevent casual operators 'skimming the cream' from established operators at peak periods. In return for this the latter are expected to provide regular services even though some of the runs involved may be unremunerative. This 'peak period' argument requires careful scrutiny though. High load factors certainly help to reduce costs per passenger mile but where the reservation of all the peak traffic to a single operator necessitates the provision of extra vehicles, which are not used at other times of day or periods of the year, the overall average load factor may not be improved at all. Indeed it is quite possible that by allowing casual operators to provide services in conjunction with the main operator at peak periods the same service could be provided with less total capacity, to the advantage of both the casual and the regular operators. Some companies effectively operate in this way by hiring coaches from contract operators at peak periods.

The precedents concerning the protection of other forms of

---

[1] See D. N. Chester, *The Control of Public Road Passenger Transport*, Manchester, 1937.

transport are a little confused. Trams and trolleybuses have usually been treated in the same way as other bus services though the Commissioners have not allowed any undertaking to regard protection as a right if the service provided was inadequate or unsatisfactory.

Though the Act enjoined the Commissioners to consider the co-ordination of all forms of transport and suggested that fares could be fixed so as to prevent wasteful competition the Commissioners have been loth to accept this invitation to define wasteful competition. A series of decisions on appeal established the general rule that road operators were entitled to a licence for any service that they could maintain throughout the year or for any seasonal service where new traffic had been created. They were not, however, allowed to enter seasonally to skim the cream from a railway service which was in operation all the year.

The third principle is that of *public need*. In the early days this was exemplified by the refusal of licences where the Commissioners were not convinced that the prospective service would be profitable and did not deem it necessary unless it could be provided on an economic basis. More recently, however, objection has not usually been raised to the running of unremunerative services so long as neither safety requirements nor alternative services were endangered. Indeed it has been one of the main aims of the licensing system to support such unremunerative services by cross-subsidization. 'Public need' has thus been interpreted not in terms of effective demand for individual services but in terms of securing the best possible network of services in an area, irrespective of the profitability of the component parts of the network.

One might sensibly ask why any operator should try to obtain a licence for an unprofitable service in any case. The answer seems to be that in the early days of licensing it was felt necessary to convince the Commissioners of one's responsibility in being willing to accept such burdens in order to be granted other, more profitable, licences. More recently such services would be provided simply to prevent the incursions of other operators on a regional monopoly. Such complete monopolies in the provision of stage services are common. The operator becomes well known to the Commissioners and they to him. The support of unremunerative services becomes a tacit link whereby the operator can guarantee the goodwill of the Commissioners.

Fares
The determination of fares is one of the most delicate of the Commissioners' duties. The Act of 1930 required that 'fares shall not be

unreasonable' and 'where desirable in the public interest fares shall be fixed so as to prevent wasteful competition with alternative forms of transport'. In the conditions then current these clauses clearly related to the danger of destructive undercutting of rates. General price stability throughout the 'thirties ensured that the problem of fare setting did not arise in any other form. Since the war, however, with perpetually rising costs, operators are frequently forced to approach the Traffic Commissioners to seek fares increases. In some areas, South Wales in particular, local authorities automatically oppose such applications.

With the prevalence of cross subsidy the Commissioners are unable to insist on fares being related to the cost of individual services. Thus they are faced with the problem of deciding what constitutes a 'reasonable' level for fares. The obvious way to do this is to authorize such fares as are necessary to protect a 'normal' rate of profit for the operating companies. Two major difficulties arise here. Firstly, if the level of profit is artificially stabilized by the licensing system then the incentive to operating efficiency may disappear. Secondly, the definition of the 'normal' rate of profit is fraught with difficulties.[1] It has been argued, for instance, that the rate of return on companies acquired at any time by the B.T.C. should be lower than that on other companies, because, in the process of acquisition, payment was made for goodwill for which there is no similar element in the nominal capital value of other companies.[2] A 5 per cent return on the nominal capital of a B.T.C. owned company may be the equivalent of a 9 or 10 per cent return on other companies if both are translated into terms of the rate of return on the current value of the physical assets involved.

Most applications for fares increases are consequent upon the rise of costs. In these circumstances it might be possible to allow only such increases as are sufficient to counteract the increases in cost. But this could only be achieved satisfactorily as the result of much fuller financial investigations than the Traffic Commissioners are able to carry out. If the present system of control is to continue it might prove beneficial to have a central investigating body to consider fares applications which could develop expertize in this complicated aspect of control.[3]

[1] See the appeal decision on a Devon General fares application in February 1963, reported in *Bus and Coach*, April 1963, pp. 146–8.

[2] *Bus and Coach*, August 1961.

[3] For an alternative suggestion see 'Granting fares freedom', *Bus and Coach*, July 1963, p. 271.

THE MARKET FOR PUBLIC PASSENGER TRANSPORT

We may now consider how road passenger services fit into the national transport system.[1] The demand for these services can be broken down into four distinct types of market, namely:

(i) Urban networks.
(ii) Long distance lines.
(iii) Country services.
(iv) Contract services, excursions and tours.

(i) *Urban networks*

Except in the case of London and one or two of the larger conurbations such as the Clyde Valley, South Lancashire and the Yorkshire woollen district, road transport is completely predominant for passenger services in urban areas.

Roughly two-thirds of London Transport receipts are derived from road services and if we assume that the takings of British Railways suburban services are about the same as those of London Transport railways we obtain an estimate that public passenger transport in London is roughly equally shared by road and rail services.

Outside London urban services are provided to a very large extent by municipal bus undertakings. Birmingham, Glasgow, Liverpool and Manchester have municipal systems with over 1,000 vehicles each whilst there are twenty-three other municipalities with between 200 and 1,000 buses and trolleybuses. In addition there are seventy-one smaller municipal undertakings and many non-municipal companies of varying sizes providing essentially urban services.[2]

There are several distinguishing features of urban passenger transport. Generally it consists of a network of radial services connecting the residential areas with the town centre, sometimes with one or two circular services skirting the area. Routes tend to be short and hence the terminal costs per route mile tend to be heavy. Frequent stops, together with urban congestion, similarly tend to produce high running costs per vehicle mile.

It has become an established principle of urban transport networks that they provide a comprehensive service. Consequently, despite the 'peakiness' of demand, they have to provide services at all times of the day and from all areas of the city irrespective of the profitability

[1] See Sir Reginald Wilson, 'The Framework of Public Transport', *Journal of the Institute of Transport*, Vol. 25, No. 5, July 1953.
[2] See *Passenger Transport Year Book*, 1963.

of individual services. Loading and therefore cost per passenger mile varies from service to service yet it is not deemed desirable to reflect these variations in differential charges. Indeed there is sometimes a taper on fares which makes price per mile lower for the longer routes where load factors are often lower and costs per passenger mile higher.

Generally this kind of comprehensive network can only be obtained by the creation of a local monopoly, usually in the hands of the municipality, to make cross-subsidization possible. Various alternatives have been suggested in an effort to introduce a competitive element into this essentially monopolist environment. For instance, the best routes could be put to auction and the profits obtained used to subsidize unremunerative services. Or a selling monopoly only could be authorized, the selling company obtaining buses and crews on charter from smaller operating units. In neither case is it likely that the resulting benefits would be sufficient to compensate for the clumsiness of the method. Where the monopoly is held by the municipality the area within which cross-subsidization has to occur is relatively small and the interdependence of the services provided may be sufficient to countervail the argument for the reduction of the less profitable services. The desires of the municipality to maintain low rates, low bus fares and a high standard of the public transport service will yield a conflict, the resolution of which lies ultimately, though indirectly, in the hands of those who both consume the service and pay the bill.

It has been argued that cross subsidy in public road passenger transport ensured by protected monopoly can fall foul of two competitors, railways and private cars. Competition from urban railway services is not usually important because they do not generally cater for the very short journeys and because the greater capacity of the individual unit of rail service usually means a less frequent and hence less convenient service. It is interesting to note, however, that in London, where neither of these defects seriously hinders the rail system, competition between bus and train has long been extinct. Both are necessary to meet peak demands. It was therefore considered undesirable that, by competition for off peak traffic, either form of transport be forced to curtail its total capacity. So long as the railways provide peak period capacity without seriously undermining bus receipts at other times of the day they must be regarded as useful and economic complements to urban bus services.

The private car provides more serious competition for the bus. In some cases (e.g. London), physical difficulties, such as the unpleasantness of driving and the difficulty of parking, may limit such competition. Because of the much larger amount of roadspace per person required by private transport any large transfer from public to private conveyance in urban areas would necessitate a massive assault, not only on the parking problem but also on the problem of the extent and nature of the urban road system. In this field the reports of the Buchanan and Crowther groups on traffic in urban areas have given some useful indication of the changes that would be required. Some cities have accepted the responsibility to ease congestion by the provision of adequate parking facilities and have even managed to provide indoor parking on a large scale at a profit. This is not a complete solution; the provision of adequate parking facilities, the costs of which the motorist is willing to meet, may serve only to accentuate the inadequacy of the road system. But there is a joint demand for roads and parking facilities. To charge an economic price for one is of no particular significance if the other is provided free. Indeed, if the motorist were to be charged an economic price for his use of scarce road space his joint demand would probably fall, making the provision of expensive parking facilities less necessary and less profitable.

(ii) *Long distance lines*

In this market road transport provides only about a quarter of the total passenger mileage travelled. It consists of lines of communication rather than an integrated national network, though the main combines do attempt to provide connecting links.

The cost per seat mile in long distance transport tends to be less than in urban transport because speed and continual running reduce both labour and fuel costs per vehicle mile. This difference is accentuated when we consider cost per passenger mile due to the extremely high load factors (up to 80 per cent) achieved on trunk routes. This is largely a result of the licensing system which limits the capacity provided on each route. Though the Traffic Commissioners can require that a regular service is provided throughout the day or throughout the year as a condition for the grant of a licence, nevertheless about half of the long distance coach services run only in the summer and about 20 per cent at summer weekends only.

The effectiveness of coach operators in competition with the railways for long distance passenger transport will depend on three main factors, namely the relative cost, speed and convenience of the alternative forms of travel. Assuming the agencies to be of equal convenience, rail transport has a general advantage of speed while road charges are normally the lower (see Table 6).

TABLE 6

Long distance passenger transport by road and rail.
Relative timings and fares, August 1963

	Approx. distance (miles)	Fastest time (hours)			Rail/road fares ratios			
Route		Road	Rail	Differ-ence	Ord. single	Ord. return	Cheap return	
London to—								
Brighton	50	2	1	1	1·50	1·65	1·55	
Portsmouth	75	2½	1½	1	1·61	1·64	1·25	
Birmingham (via M1)	110	3	2	1	1·56	1·85	·97	
Birmingham (slower route)	110	5¼	2	3¼	1·82	2·22	1·20	
Nottingham	125	4½	2	2½	1·85	2·32	1·42	(·92)
Manchester	185	8	3½	4½	2·14	2·65	1·22	
Newcastle	270	11	4	7	2·10	2·48	1·30	

Note: The rail times considered do not include those of trains on which a special surcharge is payable. Fares ratios are based on second-class rail fares and weekday road fares (weekend road fares are from 10–25 per cent above these). The cheapest return fares by rail are normally day return tickets; for Manchester and Newcastle they are overnight returns; the bracketed figure for Nottingham refers to a half day return available only on Thursdays and Saturdays during the summer.

Several interesting points arise here. Whereas the ordinary single fare by rail is based on a standard mileage rate, coach fares are considerably tapered so that the longer the journey the greater the ratio of the standard rail to the standard road fare. Similarly, the rail return fare is twice the single fare whilst the road return is usually less than double the single fare. This has the effect of compensating for the greater time difference between road and rail travel as the total journey length increases. This point may be emphasized by reference to the alternative road services to Birmingham, the faster service, being timed much more competitively with rail, is also priced nearer the rail fare than the slower road service.

The demand for long distance travel is not homogeneous in character however. The relative emphasis placed on time and on money savings will depend on the valuation which the traveller makes of his own time. Thus the business traveller may well be more susceptible to competitive timing than to competitive pricing whilst the leisure traveller, to whom the journey itself may be part of his entertainment, is more interested in the fare differential. Because of this it is possible for the railways to discriminate, offering cheaper fares to those willing to use slower or less convenient rail services. In this way they are able to compete for the type of traveller who might be tempted to go by coach without having to reduce the charge to those travellers, to whom the rail speed advantage is crucial. Consequently we find that the ratio of the cheapest return fares offered by rail and road much closer together than the standard fares. Moreover the railway 'fighting fare' is pruned most in those cases where road transport timing permits day return journeys to be made with a reasonable time at the destination. In some cases, particularly in services between provincial towns, where business travel is deemed to less important and hence competitive pricing of relatively greater significance, the day return fare by rail effectively replaces the ordinary rail rate as the competing fare. For instance, between Nottingham and Leicester the following rates prevail (August 1963):

	Ordinary single		Ordinary return		Day return	
	s	d	s	d	s	d
Road	3	8	6	3	5	4
Rail	6	0	12	0	5	3

Thus the railways, though showing a standard return fare nearly twice that of the coach in this case, have a 'fighting fare' for day return trips less than the equivalent bus fare.

Nevertheless, despite the existence of these 'fighting fares', there are many passengers, particularly holiday travellers, for whom long distance coach services have such a significant price advantage as to outweigh their slower timing.

In view of the low terminal and running costs and the absence of cross-subsidization one might have expected to find long distance coach services predominantly provided by small operators. This is not the case. The advantages of interconnection and through traffic arrangements has led to the creation of large operating groups of which Associated Motorways is the leading example. Several operating companies are involved in each of these associations. Receipts are pooled and allocated, usually in proportion to the mileage operated. Through working agreements with operators in adjacent areas make these cartels more closely akin to the four rail companies of the inter-war period, overlapping in some areas but generally content with the extent of their individual monopoly, than to the competitive market that one might imagine to exist by virtue of the number of separate operating companies in existence. Many smaller operators are involved in express services of a seasonal nature running direct from their base area to one or a number of seaside resorts. This has the effect of reducing the peak burden of the major operators, but not seriously subtracting from their regular trunk traffic.

For the regular, all year round traffic the Commissioners seem to prefer the offerings of the larger companies and associations in this field to those of the smaller operators so that, though there is a state of duopolistic competition between road and rail, there is not a great deal of competition within the road transport sector.

(iii) *Country services*

Road transport is predominant in this market. In 1961 it provided about 75 per cent of passenger mileage and, despite the current contraction of road services, the continuance of rail closures seems likely to increase this proportion, however meagre the alternative road services provided. The rural railways do not in any case form such an extensive network as the road system so that for much of their traffic in country areas the buses have an effective monopoly of public transport.

Both the average route length and the average passenger journey on country services are short. Costs per vehicle mile thus tend to be high whilst demand is patchy. Therefore, in order to support a comprehensive network of services the Traffic Commissioners have been forced to encourage cross subsidy between routes. The most convenient way of achieving this has again been by the creation of territorial monopolies within which the operator is granted sole running powers on the profitable routes as a *quid pro quo* for the maintenance of services on unremunerative routes. Though these are not always complete monopolies, in that individual routes within the territory may be run by small independent operators, nevertheless the effect of the system is that a large proportion of country services are provided by one of the larger companies able to provide enough services to ensure the necessary range for cross-subsidization.

The decline of rural bus transport. After the introduction of the new licensing system the number of bus operators declined in the 1930s whilst both vehicle mileage and passenger mileage increased. This trend continued after the war due to restrictions on private motoring. The industry ceased to expand round about 1950 and after 1955 began to decline fairly rapidly, a decrease of 10 per cent in passenger journeys occurring in the next four years. The Jack Committee on Rural Bus Services estimated that the decline in rural traffic began as early as 1952 and has been continuing at a rate of 3·4 per cent per annum since then. It attributed this mainly to the growth of private transport but also in part to the decline in evening traffic consequent on the spread of television and to the depopulation of the more remote rural areas in favour of the larger villages and small towns.

The fall in traffic led to a decline in standard of service, mainly by way of a reduction in frequencies rather than by complete abandonment of routes. The Jack Committee suggested that such hardship was involved for a certain number of people, albeit a small number, that steps should be taken to improve some services and to ensure the continuation of others. Amongst the possible remedies considered was the amalgamation of bus and postal services on the Swiss or German pattern but this was not felt to be a practical solution in British conditions. Hence some direct financial aid appeared to be desirable. Remission of fuel tax would be a simple way to grant relief but if extended to all services would be imprecise and indiscriminate, involving windfall gains for some operators who were not bearing a fair share of the burden of unprofitable routes. Partial relief of the same kind to rural services or to unremunerative services would give rise to severe administrative difficulty concerning the definition of such services. Thus it was concluded that the best way in which assistance could be given would be by direct financial aid administered by local authorities. The danger of such a system, pointed out by Mr W. James in a minority report, is that it would completely undermine the system of cross-subsidization built up over a long period by the Traffic Commissioners. In practice, the loss might not be as serious as Mr James infers, the continued decline of rural bus services having already demonstrated the insufficiency of internal cross subsidy as a method of supporting unremunerative services when conditions are so seriously adverse to public transport.[1]

(iv) *Private hire, excursions and tours*
This last category contains the irregular or unscheduled services and can be divided into two parts:

(a) Excursions and tours offered to the public require road service licences which are only obtainable so long as the service does not threaten or directly compete with any regular stage or express. Moreover stringent conditions are usually attached to this type of licence to protect existing services. Consequently the extent of these types of service is very limited and in 1961 they were only responsible for 3·5 per cent of total road passenger transport receipts.

[1] Both the Ministry of Transport Highland Transport Enquiry and the Transport Committee of the Council for Wales and Monmouthshire have subsequently reported in a similar vein, though in both cases declaring that any necessary subsidy should be financed by the Exchequer.

(b) Private hire of coaches can be carried on without the need for a road service licence and is therefore the one corner of the road passenger transport industry where real competition prevails. Moreover, though some of the larger firms have specialized luxury vehicles for this trade, it is predominantly the field of small scale operation. The small operator, of whom there are over 3,500 operating five vehicles or less, may utilize the intense industry of the owner driver and spare time drivers to provide a service of flexibility and economy. In fact, a very large proportion of the receipts of the smaller operators is obtained by this type of work (see Figure 9).

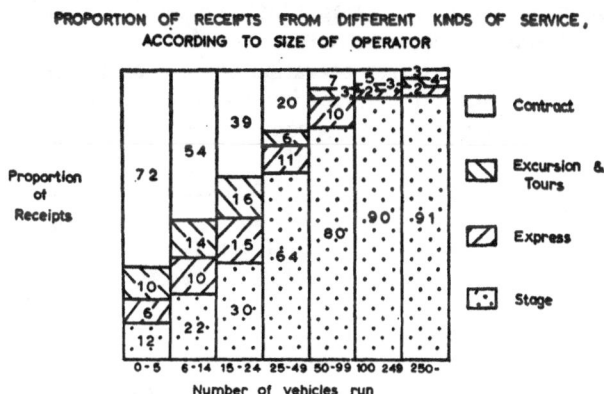

PROPORTION OF RECEIPTS FROM DIFFERENT KINDS OF SERVICE, ACCORDING TO SIZE OF OPERATOR

Source

Based on Ministry of Transport Public Road Passenger Transport Stastistics 1961-2

Figure 9

In many cases small operators are only able to maintain un-remunerative rural stage services by virtue of the profits obtainable from contract work, either regular, in the form of schools or works contracts, or irregular, in the form of private outings. So long as the private hire is genuine and a separate demand is catered for, there seems to be no reason why the freedom allowed in this sector should be curtailed. One of the major problems produced by the licensing regulations is to decide whether a service lies genuinely in the category of private hire. The provisions of the Road Traffic Acts on this point are complex and the decisions of the courts are necessarily based on technical points of this complex law. Attempts by the small operator with no licensed services to extend his business in this field may therefore prove hazardous from a legal point of view. In this respect the larger operator who has already established his status with the Traffic Commissioners is at an advantage, for, if he is in any doubt as to the legality of an unlicensed hiring, he may play safe by applying for an excursion licence.

THE RESTRAINTS ON COMPETITION

There are nearly 350 operating companies with twenty or more vehicles and over 4,000 smaller operators engaged in road passenger transport. Only ten concerns operate more than 1,000 vehicles, the largest of which, Birmingham and Midland, with 1,900 vehicles, accounts for less than 3 per cent of the total. The relatively small scale of operation does not imply that competition is stifled solely by the licensing system, for two reasons.

Firstly, few of the larger operating companies are independent. Over half of the industry is municipally or State owned; 15 per cent is controlled by a private holding company, British Electric Traction, though the State owned Transport Holding Company also has an equal interest with B.E.T. in many companies in the group. This leaves only 32 per cent, composed mostly of small operators, in genuinely independent hands (see Table 7). Linked control must inevitably take the edge off competition despite apparent independence of operation.

TABLE 7

Ownership of Road Passenger Transport Vehicles, December 31, 1961

		Number of vehicles	% of total
Municipal		18,256	23·8
State, of which		22,513	28·9
London Transport	8,366		10·7
B.T.C.	14,147		18·2
Other operators, of which		36,738	47·2
B.E.T.	11,540		14·8
Independents	25,198		32·4
Total		77,777	100·0

Sources: Public Road Passenger Transport Statistics, 1961–2, B.E.T. figure from Passenger Transport Year Book, 1963.

Secondly, there is collusion of a more explicit form. The cartelization in long distance coach services which we have already mentioned crosses the boundaries even of group ownership. The country is also divided territorially by what are known as 'area agreements'. These are bilateral agreements between large operating companies in contiguous areas stating the boundaries over which they will not trespass except by explicit consent. Despite the fact that the legal status of these agreements is not very clear, and that there may be smaller operating companies competing for licences within the agreed areas, they nevertheless prevent competition between the larger operators. Because the agreement relates to a service it does not qualify for registration under the Monopolies and Restrictive Practices Act, 1956, so that the only public control over this type of practice is that exercised by the Traffic Commissioners. In so far as fares lie within the jurisdiction of the Commissioners, 'exploitation' in the form of high profits is unlikely. But this is not the only danger. One of the most welcome of monopoly profits is a quiet life; the present system does little to subject existing operators to any competitive spur.

The Traffic Commissioner is a kind of planner; like other animals of the species he has a natural predisposition towards tidiness and convenience which is best achieved when dealing with established and responsible operators. A ministerial directive in the early 'thirties required that special attention be paid to the interests of the very small operator who thus also receives fairly generous treatment. There is a danger however that the medium-sized operator is the one to be squeezed out. In order to obtain licences he must offer a high standard of service and facilities, involving heavy overhead costs, before he has been given the chance to obtain the traffic over which to spread these costs. In these circumstances it is almost impossible to challenge existing operators, however inefficient, in the field of licensed services.

Public road passenger transport competes with both the railways and with private motoring but only in the small corner of unlicensed services is there any real competition within the industry itself.

CHAPTER EIGHT

Road Haulage

Road haulage is responsible for a large, and increasing, proportion of freight transport in Great Britain. Total road ton mileage has increased by about 55 per cent and the proportion of inland freight haulage accounted for by road has increased by nearly 20 per cent over the last decade (see Table 8).

TABLE 8

Inland Freight Transport 1952–62

	Road	Rail	Total	Road as % of total
	('000 million ton-miles)			
1952(a)	18·8	22·4	41·2	45·6
1958(a)	23·1	18·3	41·4	55·8
1962	29·1(b)	16·1(c)	45·2	64·4

Sources: (a) 'The Transport of Goods by Road', H.M.S.O., 1959. Further interpolations for period 1952–8 can be found in K. F. Glover, 'Statistics of the Transport of Goods by Road', *Journal of the Royal Statistical Society*, Series A, Vol. 123, 1959.

(b) Based on 'Survey of Road Goods Transport, 1962, First Results', Ministry of Transport, March 1963.

(c) B.T.C. Transport Statistics, 1962.

Such a simple comparison inevitably oversimplifies matters. It appears, from the crude figures, that there has been a huge transfer of traffic from rail to road, suggesting a clear overall superiority of road transport. In fact, road and rail transport typically perform different types of work. The average rail haul is about 70 miles as opposed to only 25 miles for road haulage. Rail freight is generated very largely by the coal and steel industries whilst road transport is concentrated much more on the distribution of consumer goods of all kinds. It is the decline in the former industries and the advance of the latter which largely explain the changes between 1958 and 1962. The changing emphasis in favour of road transport thus appears to stem as much from the secular decline of those activities depending on rail transport and advance of those depending on road as from actual switches of traffic from rail to road.

Even within road haulage the crude aggregates hide more than they reveal. There are wide variations of vehicle type and size, fleet size, method of operation and so on, reflecting both the nature of the traffic offering and the effects of the statutory controls to which the industry is subject. The most basic distinction of all is that between public and private haulage, enforced by the licensing system.

THE LICENSING SYSTEM

This licensing system was a product of depression. Even before 1930 there were signs that the reorganized transport framework resulting from the 1921 Railways Act was not entirely satisfactory; the slump convinced the Government that further intervention was necessary.

The Select Committee set up in 1929 to investigate the workings of the transport system was far less successful in dealing with freight than it had been with passenger traffic. Indeed it was so devoid of concrete suggestions that the problem was passed back to the industry itself in the form of the Salter Conference in 1932. This conference was composed of four representatives of rail and four of road interests under an independent chairman. It was to consider and to make recommendations as to how a fair basis of competition could be achieved between road and rail.

The Conference saw the need for three major alterations in the form of control of road haulage.[1] In the first place, it advocated a redistribution of the burden of road expenditure, the general effect of which was to cast more of the responsibility on to heavy vehicles. Henceforth, it was argued, road hauliers would meet their full share of track costs and were thus placed in a position of fair comparability with the railways in this respect.

The second suggestion was that road hauliers should be subject to some control as to wages and conditions of service, both on welfare grounds and to put them on the same footing as the railways. A distinction was made here between the public haulier who was to be subject to public control, and the ancillary haulier, who was exempt from such control.

Thirdly, this distinction was put to further use. In view of the arguments of the Road Haulage Association that the troubles of the industry were largely attributable to excessive competition, it was suggested that, by means of a licensing system, the ancillary user be prevented from competing with the public haulier for any traffic other than that carried on his own account and that public haulage capacity also be controlled in the public interest.

The recommendations of the Salter Conference were implemented

[1] See the *Report of the Conference on Rail and Road Transport*, July 1932.

I

in detail by the first part of the Road and Rail Traffic Act, 1933, and were recently consolidated in the 1960 Road Traffic Act. All vehicles used for haulage, other than farmers' and Government owned goods vehicles, are required to hold a carriers' licence. Carriers' licences are of four types:

'A' licences are for the general carriage of goods for hire or reward. 'Contract A' licences are for the exclusive carriage of another person's goods under a contract for a continuous period of not less than a year.
'B' licences are for the carriage of the trader's own goods, or subject to the special provisions of the licence, for hire or reward.[1]
'C' licences are specifically for the carriage of goods in connection with the business of the licensee.

There are eleven licensing areas corresponding to the Traffic Areas for road passenger transport and the Licensing Authority for each area is the chairman of the Traffic Commissioners for the area. Hence control of goods and passenger transport licensing for any area is in the same hands. The grant of a 'C' licence to carry goods on own account is a right subject only to safety provisions whereas in the case of both 'A' and 'B' licences the grant lies at the discretion of the Licensing Authority.

In exercising their discretion the Licensing Authorities were to pay regard to the interests of those providing as well as of those requiring transport facilities. This was interpreted to mean that before any new 'A' or 'B' licence be granted it should be demonstrated both that there was need for a service and that the existing capacity was not adequate to meet this need. The Transport Act of 1953 amended this slightly. Firstly, it gave precedence to the interests of those requiring transport instead of giving them equal consideration with the providers of facilities. Secondly, it enacted that the onus should be on the objector to a licence application to prove the absence of need for the extra capacity rather than on the applicant to prove the need. Thus the onus is on existing operators to prove a case against new entrants. Thirdly, new entry was also facilitated by the provision that the prices charged or to be charged were henceforth to be relevant evidence on a licence application. Thus price competition became possible at the licensing stage.

In contrast with these changes, all of which would seem to imply greater freedom of entry and competition, there is one contrary trend.

[1] In fact over three-quarters of the work performed by 'B' licence vehicles is now for hire or reward so they can be viewed as predominantly public haulage. See 'Survey of Road Goods Transport, 1962, First Results', Table 15.

All licence applications must contain a statement of the type of traffic and area of operation for which the licence is required. This is known as the 'normal user'. The more vaguely it is stated (e.g. general goods, Great Britain) the greater the likelihood of objections to the grant. The Licensing Authorities have the power to revoke, suspend or refuse to renew licences on the grounds that this statement of intention has not been fulfilled. In the last decade a tighter control appears to have been exercised in this respect. Thus even the 'A' licence, unless it is for 'general goods, Great Britain', does not constitute unrestricted freedom for the haulier. The chief exception to this detailed control is the 'A contract' licence which is automatically granted if the applicant can show that his vehicle will be used exclusively for the purpose of a contract for the carriage of another trader's goods.

The recommendation of the Salter Conference that entry into the road haulage industry be rigorously controlled was based on three principal grounds. Firstly it was argued that cut-throat competition in freight rates was inimical to the continued existence of a stable road transport industry. In particular it was said that the effects of ancillary hauliers and small public hauliers obtaining return loads from clearing houses, sometimes as the result of a 'Dutch auction', made a stable freight structure impossible to maintain. As a result of this many freights were carried at less than their full cost and bankruptcies resulted.

The second argument followed from this. It was asserted that, in face of this intense competition, maintenance was being neglected and owner drivers were forced to work dangerously long hours. Both of these factors were likely to result in accidents.

Thirdly, in the generally depressed and unstable conditions of trade at the time it was believed that the transport industry as a whole would benefit from 'co-ordination' between road and rail. If a more rigorous control of road haulage was the price at which this could be achieved then many thought it best to pay that price, though in fact even the Salter Conference was unable to recommend any specific measures to guarantee 'co-ordination'—or even to define it.

In view of the magnitude of the changes which were to depend on these arguments it is surprising that they should have been subject to so little public scrutiny at the time. The Salter Conference leant heavily on the evidence of the Road Haulage Association, then a very small body indeed. The R.H.A. was the successor to the Long Distance Road Haulage Committee, a body consisting of representatives from a few of the larger hauliers convened solely for the purpose of presenting a case to the Royal Commission.[1] It naturally advocated a solution beneficial to the larger road haulage companies irrespective of the effects on the smaller man. Despite its unrepresentative nature and the fact that its view differed radically from those previously expressed to the Royal Commission by the Commercial Motor Users Association, the R.H.A. succeeded in securing a largely uncritical acceptance.

Let us examine the case more closely. In view of the fact that the evidence was given in the middle of the worst slump of the century it is not surprising that transport rates were unstable; indeed it would have been much more surprising if they had not been. If the clearing house system, and particularly its alleged abuse, was an important contributor to this instability then the obvious solution would seem to be control of the clearing houses rather than restriction of the whole industry. No statistics were produced to support the allegation that the industry was unduly prone to bankruptcies and a recent investigation[2] has undermined this proposition. Road haulage, even before the Act, had a lower rate of bankruptcy than most of the other trades where small scale operation was equally dominant. Though the number of bankruptcies in road haulage declined after the introduction of the Act this was inevitable by the very nature of the controls involved and does not prove that the original rate was unduly high.

A similar criticism can be applied to the argument concerning accidents. Statistics for street accidents, classified by cause, are available since 1927. An analysis of these accidents[3] shows that the accident rate per '000 vehicles was slightly higher for goods vehicles than for private cars but only one-third that for buses and coaches. The rate for goods vehicles improves during the 'thirties but so does that for private cars so that this can probably be attributed to a very large extent to the general improvements in road safety after the 1930 Road Traffic Act. Moreover, in the analysis of fatal accidents, there is a little substantiation of the 'owner driver' argument. The proportion of accidents attributed to inadequate maintenance or driver fatigue is insignificant and gives more support to the C.M.U.A. assertion before the Royal Commission that goods vehicles in commercial ownership were, in general, as well maintained as private vehicles, than to the R.H.A. claim that small scale operation was indirectly responsible for a serious threat to public safety.

[1] P. E. Hart, 'The Restriction of Road Haulage', *Scottish Journal of Political Economy*, Vol. 6, No. 2, June 1959.
[2] W. M. Macleod and A. A. Walters, 'A Note on Bankruptcy in Road Haulage', *J. Ind. Econ.* Vol. V, No. 1, November 1956.   [3] P. E. Hart, *op. cit.*

Finally the 'co-ordination' argument. This bore weight because of its vagueness; it was indeed 'all things to all men'. To the railways it meant the power to prevent extension of the road competition by means of objections to 'A' licence applications, to the larger haulier it meant the end of undercutting of their rates by smaller operators; to the trades unions involved it meant a more stable framework within which to achieve improvements of wages and conditions; to the radical left it meant a step nearer to nationalization without which they considered that perfect co-ordination was impossible. Only to the small haulier, of the interested parties, did it spell an undesirable and artificial introduction of rigidity into an industry the nature of which required flexibility.

With the exception of one or two minor modifications ('A Contract', 'Special A' and 'C' hiring allowances), and despite being largely suspended during the period of nationalization of road haulage, the system has remained essentially unchanged for over thirty years.

THE STRUCTURE OF THE ROAD HAULAGE INDUSTRY

We must now examine the structure that has emerged from this form of control (see Table 9).

TABLE 9

Road Vehicle Licences, 1938–61

| Category | Number of licences ('000) | | | | |
	1938	1948	1953	1958	1961
'A' (and 'Special A')	83·7	74·2	89·3	88·5	88·0
'Contract A'	9·5	16·3	9·5	20·4	27·4
'B'	54·9	65·5	58·8	64·7	69·8
'C'	365·0	590·5	866·3	1,099·3	1,253·8
Total	513·1	768·2	1,023·9	1,272·9	1,439·0

Source: Annual Abstract of Statistics, 1962.
Notes: 1938 figures are for the end of June, all subsequent figures for the end of the year. In 1948 there were 21,600 B.T.C. vehicles not included in other categories.

The most notable feature is the rapid expansion of the number of 'C' licences after the war compared with the relatively slow increase in other categories. But the vehicles within each category vary both in type (Table 14 at end of chapter) and size (Table 15). Platform or sided vehicles are most commonly used for public haulage or for heavy 'C' licence work. Box-bodied vehicles predominate amongst the smaller 'C' licensed vehicles. As for size, it is noticeable that while over 70 per cent of 'C' licensed vehicles have a carrying capacity of not over 2 tons only 9 per cent of 'A' licensed vehicles were as small as this (and of these B.T.C. collection and delivery vehicles probably accounted for nearly half).

The suggestion of different types of work implicit in these figures is borne out by other statistics. For instance, whereas 62 per cent of the total 'A' licence ton mileage is performed on hauls of over 100 miles, only 22 per cent of 'C' licence ton mileage falls in this category. Moreover, whereas over 70 per cent of 'A' licence mileage was involved in end-to-end work only 30 per cent of 'C' licence mileage was of this type. As we would expect then, an analysis of 'C' licence vehicles (Table 16) indicates that over 30 per cent are engaged on retail delivery work and that this proportion is highest amongst the smaller vehicles.

These characteristics of the road haulage industry in 1962 may be summarized neatly in the following way (see Table 10).

TABLE 10

Road Goods Transport during the week April 2 to 8, 1962

Licence Category	Licences		Tons carried		Ton mileage	
	'000	%	millions	%	millions	%
'A'	89	6·4	4·1	16·4	198	29·4
'Contract A'	29	2·1	1·9	7·6	67	9·9
'B'	72	5·2	4·3	17·2	71	10·5
'C'	1,206	86·4	14·7	70·8	338	50·2

Source: Figures from 'Survey of Road Goods Transport, 1962, First Results'.

'C' licences or private haulage accounted for 86·4 per cent of vehicles licensed. But within this category were a disproportionate number of the smaller vehicles so that in terms of tonnage carried the private haulier was only responsible for 70·8 per cent of the total. But the smaller vehicles, by virtue of the work they were doing, ran comparatively small weekly mileages at relatively low load factors. Consequently in terms of ton mileage, the work done by the 'C' licensed vehicles accounted for little more than 50 per cent of the total.

THE EFFICIENCY OF THE PRIVATE CARRIER

The varying size distribution of vehicles is not a sufficient explanation of the disparity between the amount of work done per vehicle under 'A' and 'C' licences for similar disparities occur in each individual weight category. Moreover, the same pattern was revealed by a similar survey carried out in 1958.[1] It is very tempting therefore to

[1] *The Transport of Goods by Road*, H.M.S.O., 1959.

conclude that the 'A' licensee makes more efficient use of capacity than the operator on own account.

In 1958 the organization of the 'C' licence operators, the Traders Road Transport Association, carried out their own investigation which, they claimed, refuted this allegation by demonstrating the absence of serious empty running on 'C' licences.[1] In their survey, which covered 9 per cent of all 'C' licence vehicles, only 17 per cent returned empty, a figure which could be reduced to only 8 per cent if we exclude such vehicles as bulk liquid carriers and tippers which do not usually carry return loads. But the majority of the vehicles in the T.R.T.A. survey were fairly small; for vehicles over 5 tons, however, about 40 per cent returned completely empty. Moreover, the longer the journey for 'C' licence vehicles, the greater, according to the T.R.T.A. survey, the chance of an empty return run. Clearly the T.R.T.A. survey did not satisfactorily dispose of the argument that, for vehicles of comparable size, empty running is more prevalent and load factors lower for 'C' than for 'A' licence operation.

A more tenable defence of 'C' licence operation was given by the Ministry of Transport in the report on its own 1958 survey. Instead of denying the existence of the disparities in utilization of capacity it sought to interpret them. It gave four reasons why figures of utilization could not justifiably be used to support any positive comparison between the efficiency of 'A' and 'C' licence operations.

Firstly, the figures are only averages which cover considerable variations within classes. For example, although the average weekly mileage of large (over 5 tons) 'A' vehicles is greater than that of large 'C' vehicles, nevertheless 30 per cent of the 'C' vehicles in the category ran a weekly mileage greater than the average for 'A' vehicles.

Secondly, there are two different types of work in which vehicles may engage. In 'intermediate' work a vehicle may be setting down or picking up goods at several points during a journey; in 'end-to-end' work a complete load is carried from one picking up point to one destination. Vehicles tend to show higher mileages, ton mileages and load factors when engaged in 'end-to-end' working irrespective of their licence class. But the Ministry survey revealed a larger proportion of 'C' vehicles than of 'A' vehicles at all weights engaged in intermediate working with a consequent depressing effect on operating statistics.

Thirdly, the weekly mileage and ton mileage also depend on average length of haul. The lower the average length of haul in any class the lower the weekly mileage and ton mileage figures. But the average

[1] Traders Road Transport Association, *Survey of 'C' Licensed Vehicles*, October 1959.

length of haul for 'A' licensed vehicles over 5 tons was nearly twice that for 'C' licences so that the better ton mileage figures may partly represent different spheres of activity.

Finally, the comparison was affected by the fact that vehicles of the same unladen weight may differ considerably in carrying capacity. The survey revealed that 'A' licence vehicles possessed, on average, between 13 per cent and 25 per cent more carrying capacity than 'C' licence vehicles similarly classified according to unladen weight due to the greater tendency for 'C' licence vehicles to be specially fitted for particular traffics. The difference in loading thus partly represents a difference in the commodities carried.

To these four reasons we may add one more. Clearly, services are not homogeneous. The crude statistics cannot indicate fully the extent to which services met and suited the needs of the user. Therefore, to understand the belief of a trader that he can provide his own transport more efficiently than the public haulier, we must appreciate the greater variety of attributes which together constitute an 'efficient' service in his eyes.

Information of this kind was made available when, in June 1958, the Traders Road Transport Association undertook a sample survey of its membership. This random sample covered 4,837 operators running 98,340 vehicles, or 9 per cent of all 'C' licensed vehicles, a sample large and unbiased enough to be statistically respectable. In this survey operators were asked, *inter alia*, what advantages induced them to operate their own transport services in preference to the use of public facilities. A number of possibilities were suggested; the respondents were merely asked to indicate which were relevant to their own company. In Table 11 we reproduce the results, showing the percentage of all vehicles surveyed for which each reason for preferring private to public transport was deemed applicable.

The survey certainly suggests that the relative cost of transport is an important element in the decision to run 'C' licensed vehicles but it makes it quite clear that the total real cost be be considered includes allowance for such properties as speed, certainty of timing, security, publicity, etc., on which grounds traders attribute considerable importance to the possession of direct control of the means of transport. There is no indication, however, of the extent to which traders make proper comparison of the relative costs of carriage by their own vehicles and public haulage either of their total traffic or of marginal loads. A growing cost consciousness amongst ancillary hauliers might possibly cause some traffic to return to public transport.

TABLE 11

The Advantages of Transport on Own Account

Reason	% of vehicles for which private road haulage was preferable to	
	Public road haulage	Rail
Speed of delivery and certainty of timing	68	75
Cost	44	56
Placing of goods in premises beyond point normally served by public transport	32	42
Avoidance of breakage or damage	33	44
Avoidance of pilferage	21	25
Reduction in packing costs	24	31
Prompt return of empties	30	32
Special vehicles used	29	31
Services provided by driver (sales, collection, etc.)	39	32
Advertising on vehicle	37	32

Source: T.R.T.A. *Survey of 'C' Licensed Vehicles, op. cit.*

There are two further conclusions concerning the relative costs of private and public haulage which can be deduced from the T.R.T.A. survey.

Firstly, there is rather less scope for the improvement of load factors by the transfer of traffic to public haulage than at first appears to be the case. Over 50 per cent of 'C' licensed vehicles of over 5 tons unladen weight were of specialized construction or use.[1] It is these vehicles in particular which secured such poor return loadings; the more specialized the type of vehicle the greater the proportion of empty running. But the demand for such vehicles is limited not merely by the licensing restriction on securing return loads for hire but also by the unidirectional flow of the commodities concerned. Therefore, if the commodities cannot be carried in a general purpose vehicle and the specialized vehicle cannot be put to any alternative use, neither public road haulage nor rail transport would achieve better loadings than the 'C' licensee. This contention is substantiated by the breakdown of return loading according to industry. 'C' licence operators obtain their poorest results in the fuel, mining and quarrying industries where the nature of traffic (and hence of vehicle) and the unidirectional flows are such as to make return loads as unobtainable for the public hauliers as they are for the 'C' licensee.[2]

The second deduction concerns costs and fleet size. Both for short and long distance haulage the return loading factor is poorer the smaller is the 'C' licence fleet involved. But it has been made quite

[1] T.R.T.A. Survey, *op. cit.*, Table 5.
[2] T.R.T.A. Survey, *op. cit.*, Table 19.

clear by recent railway costings investigations that the railways will be able to provide particularly cheap and effective service, tailored to the needs of the customer, in cases where a large, and particularly a regular, flow of traffic is forthcoming. Hence it is likely that it will be the larger 'C' fleet, which apparently makes more effective use of capacity than the smaller fleet, which is most likely to be superseded, wholly or in part, by railway service.

There is little doubt that, in the post war period, 'C' licence operation has been stimulated by the restrictions on entry into the field of public haulage and the failure of railway commercial policy to offer low rates where profitable to obtain or keep traffic. On the other hand, despite the absence in many companies of any very sophisticated accounting system for transport costs, it is unlikely that the 'C' licensee is quite as inefficient in relative terms as his poor load factors for certain traffics and vehicle categories would seem to indicate.

THE PUBLIC HAULIER

Though 'A' licences are sometimes used for purely local carriage and even, in some instances, for urban delivery services, the public haulier is more typically engaged in medium and long distance carriage.

Despite the existence of the nationalized British Road Services, operating some 16,000 vehicles at the end of 1962, the industry is still predominantly one of small scale operation. Exactly how small it is difficult to say. The licensing figures indicate an average size of less than five vehicles per 'A' licence operator. Due to their method of collection, however, these figures may underestimate the true average size of the public haulage company; moreover, there have been amalgamations in recent years and increasing emphasis on inter-working agreements and exchange of traffic which would not show up in the licensing statistics.[1]

[1] A trend towards larger units is suggested by P. S. Henman, 'The Economics of Road Goods Transport', *J. Inst. T.*, Vol. 29, No. 3, March 1962.

There are no reliable statistics to support this point. The ratio of numbers of vehicles to number of operators in each licensing category shown in the Licensing Authorities' Annual Reports has not shown any significant trend over the last decade. Nevertheless a trend towards amalgamation might not show in the figures, which simply indicate the ratio of vehicle numbers to the number of separate licences issued. The average size of fleet is greater than that suggested by these figures because individual operators do not always have all their vehicles on one licence. After an amalgamation the operator may retain separate licences, and the statistics remain unaltered, for several reasons:

The cost structure of road haulage will, of course, vary to a certain extent according to the size of operator and the type of commodities carried and service provided. For the largest operator we are well informed. As a nationalized concern B.R.S. is required to publish accounts in considerable detail; the B.R.S. cost structure for 1962 is shown in Table 12.

TABLE 12

The Cost Structure of British Road Services, 1962

	£'000		%
Vehicle Operating expenses			
Drivers and Attendants	15,769		27·8
Fuel	5,804		10·2
Tyres	1,892		3·3
Lubricants	107		·2
Other	2,450		4·3
		26,039	46·0
Maintenance of Vehicles			
Repairs	5,456		9·6
Depreciation	3,567		6·3
Administration	314		·6
		9,336	16·5
Other Traffic Expenses			
Depot expenses	8,134		14·4
Compensation and Insurance	1,438		2·5
Other	2,077		3·7
		11,749	20·7
Maintenance of buildings	620		1·1
Vehicle licence duties	1,283		2·3
General	7,615		13·4
Total		56,642	100·0

Source: B.T.C. Annual Report and Accounts, 1962, Vol. II, pp. 81–3.

For smaller firms the overhead element will tend to be less important and, particularly in the case of short distance operator, wages may account for as much as half of the total costs. In order to spread vehicle overheads drivers are encouraged to work long hours and the fairly low basic wage for a forty-two hour week is

(a) Inertia.
(b) Geographical reasons; where an organization has depots in different cities or licensing areas it is convenient to have separate licences.
(c) Administrative reasons; many amalgamations are purely financial, the units retaining a high degree of autonomy in operation and hence holding their own licences.

normally supplemented by substantial overtime payments. Unfortunately much of this paid duty time is often wasted by loading and unloading delays. As yet there has been no universal application of detention or demurrage charges in road haulage with the result that traders may organize their concerns irrespective of any delays caused to hauliers. Good customers thus cross-subsidize bad customers to a certain extent in this respect, and from the community's point of view, there is insufficient incentive to achieve the right balance between production and transport costs.

The prevalence of loading delays has adversely affected costs in less direct ways. The risk of delay frequently makes it impossible for the trunk vehicles to be used also for local collection and delivery of the goods transported. Consequently feeder vehicles have to be employed, thus increasing haulage costs. The problem of loading delay is most serious where part loads are involved and delays tend to cumulate. Therefore, as an incentive to full loading, the Road Haulage Association Long Distance Committee has recommended a schedule of differential rates favouring vehicle load consignments.[1]

The R.H.A. recommend schedule is no more than a rough substitute, however, for charges based on the hauliers detailed knowledge of the costs of his own individual operations. Even for the large haulier variable costs constitute a large proportion of the total and the divisibility of the capital equipment (most of which may consist of his vehicles) may enable him, without too many problems concerning the allocation of joint overheads, to assess fairly precisely the cost of using a particular vehicle for a particular run taking a particular period of time.

If standing charges are allocated between vehicles according to capacity, drivers' wages and expenses expressed per vehicle week and running costs expressed per vehicle mile, then for a standard mileage we can obtain a weekly standard costing for each vehicle, an example of which is shown in its simplest form below.[2]

The preparation of these budgets will entail a certain amount of

Weekly Costs of a 10 Ton Capacity Vehicle

	£	s	d
Driver's wages and expenses	19	17	4
Running costs—620 miles at 8½d per mile	21	19	2
Standing charge (10 ton at 8s 3d per ton per day)	20	12	6
	62	9	0

[1] P. S. Henman, *op. cit.*, p. 269.
[2] From Henman, *op. cit.*

initial investigation and clerical work. Once prepared, however, they can be used in two important ways to ensure efficient operation:

(a) Examination of the variances from the standard costs would enable swift corrective action to be taken in the case of avoidable increases.

(b) Reference to the normal standards of performance will enable the operator to avoid quoting rates below the 'opportunity cost' of operation where competition is fierce. Where alternative traffic is not available the true 'opportunity cost' will only include the 'escapable costs' which are the running costs of the vehicle and some part, if not all, of the driver's wages and expenses. Assuming that depreciation has been allocated correctly between running cost and standing charge the opportunity cost in these circumstances would not include any contribution to standing charge.

We must not claim too much for these techniques. The fleet operator has still to secure the most profitable traffic and allocate it between his vehicles. The standard costing method is an aid to good management; it is not a substitute for it.

THE FUTURE OF THE INDUSTRY

The organization of the industry, and particularly the licensing system, appears to be coming under increasing pressure. A great deal of energy is expended in attempts to evade the limitations imposed by the system. Some of the devices, such as the application for 'A Contract' licences in the hope that they can later be converted into general 'A' licences and the use of 'C' hiring allowances, which permit lorries to be transferred from user to user on a very short term basis, are within the letter, if not the spirit, of the law. For the less scrupulous haulier there are numerous illegal evasions which can be practiced, including forged licences, bogus contracts and straightforward operation beyond the limits imposed by the licence held. During the last two or three years several Licensing Authorities have expressed concern in their annual report over the apparent increase in some forms of illegalities, particularly those relating to evasion of the licensing conditions. Though an increased number of convictions has sometimes simply reflected more thorough enforcement, nevertheless the profusion of illegal practices appears to be increasing. In 1962 there were 25,390 convictions for offences within the concern of the Licensing Authorities, of which 6,902, or 27 per cent, were for the evasion or breach of licence conditions. Moreover, most convictions represented not an isolated offence (which would probably be dealt with by a warning letter from the Authority) but a continued failure to comply with the regulations (see Table 13). Nor do these figures represent the full extent of evasion; there are a number of other misdemeanours for which the Licensing Authority may suspend, curtail or revoke licences without recourse to the courts. These include falsification of applications, infringements of normal user and persistent charging of rates below cost to secure an unfair advantage.

Enforcement of the regulations in the courts has frequently produced anomalies. Treatment of convicted offenders has been variable.[1] In some cases magistrates have criticized the practice of laying multiple charges to show that evasion was wilful and frequent, yet where single charges have been laid offences have been pardoned on the grounds that they were isolated and unwitting occurrences.

TABLE 13

Summonses issued by, or in co-operation with, the Licensing Authorites, 1961–2

Offence	Prosecutions Number	Convictions Number	%
Using a goods vehicle without a carriers' licence	3,997	3,839	15·1
Breach of licence conditions			
'Contract A'	577	575	2·3
'B'	1,401	1,373	5·4
'C' (Use for hire)	1,228	1,115	4·4
Hours and Records offences	15,772	14,831	58·4
Construction and maintenance	2,514	2,350	9·3
Other offences	1,394	1,307	5·1
Total	26,883	25,390	100·0

Source: Annual Reports of the Licensing Authorities, 1961–2.

The extent of evasion and the anomalies of enforcement suggest that the licensing system has itself fallen into disrepute, not only in the industry itself but also, to a certain extent, in the courts. In view of this, an independent inquiry into the system, under Lord Geddes, was instituted in July 1963. The principle of central regulations for safety, standards of construction and maintenance and labour conditions is not seriously in question so that the retention of some controls is not disputed. But the restrictions on the freedom of entry

---

[1] In 1960 the report of the L.A. for the Yorkshire area referred to a case where an absolute discharge was granted despite the fact that the defendant had pleaded guilty to 110 different charges of evasion or breach of licence conditions. In the same year, at the Southend Court a party was fined £575 plus costs for a similar list of offences.

into the industry may stand or fall quite independently of the safety regulations; the most difficult task of the committee will be to estimate the likely effects of changing or abolishing the present system on the efficiency of transport as a whole.

The spectre of cut-throat competition

It has often been argued that the possibility of entry into the road haulage industry on a very small scale, allied to the fact that it is better to accept any traffic paying more than prime cost rather than stand idle, will, in the absence of restrictions on entry, result in competitive rate cutting of such severity that no haulier will be able to cover his full costs. An argument of this kind was the basis for the introduction of a restrictive licensing system in 1933.

There are two good reasons why such an argument is now of diminished validity.

(a) On the supply side, such price cutting would result in very low rates of profit for the operators involved and, in the case of owner-drivers, very poor remuneration for labour. When alternative opportunities are available for profitable investment of capital and well paid employment of labour it is unlikely that a succession of small operators will enter an industry offering such poor prospects.

(b) On the demand side, stability and reliability of service is increasingly sought by traders. The expenses incurred as a result of any discontinuity in the flow of either supplies or product places a premium on the sort of assured service which it may only be possible to obtain from a fairly large scale operator, often on a fairly long term contract or agreement.[1]

Despite the apparent advantages of large scale operation in terms of reliability of service the small haulier still plays an important part in maintaining this reliability. Both the large public haulier and the ancillary haulier may at times find themselves over-committed so that an uninterrupted service can only be maintained by subcontracting to the small haulier. Such business is naturally volatile, disappearing when total transport demand contracts, so that too great a reliance on this sort of work can be disastrous for the small haulier; nevertheless, it provides a useful supplement to the work that the small man can obtain direct through his own contacts.

[1] The extension of 'A Contract' licence operation in recent years has been partly due to this trend for long term agreements and partly due to the impression that public 'A' licences would be more easily obtainable by conversion from 'A Contract' than by original application.

Nor is large scale operation without its attendant disadvantages. For instance, one of the chief reasons adduced for the denationalization of road haulage in 1953 was that B.R.S. was so large and its form of management so centralized that the standard of service suffered, particularly by the absence of personal contact and responsibility. Under the current structure of the industry, however, concentration can proceed for a long time before these same disadvantages will again be encountered.

Both by the amalgamation of companies and by inter-working agreements and the exchange of traffic the effective scale of road haulage operation appears to be currently increasing. Such a trend, together with improved knowledge and control of costs, make it unlikely that the industry would again suffer, even in a recession, any serious damage due to 'excess competition'.

Efficiency and the licensing system

Even if the Geddes Committee is similarly unsympathetic to the 'cut-throat' competition argument it will still need to assess and make recommendations concerning the effects of changing the licensing system. Implicitly, at least, the committee must define efficiency for the road haulage sector. But, as we argued earlier (in Chapter 1), efficiency in a competitive transport sector consists partly of securing low cost operation and partly of producing services suited in character to the particular needs of transport users. We shall now briefly examine each of these elements in turn.

Firstly let us consider the cost of transport for the economy as a whole. Because transport services cannot be stored we may regard the total amount of capacity existing in the industry as an indicator of the total cost of meeting the transport needs of the community as a whole. It is not clear how the abolition of the licensing system would affect this. The fact that there is at present a thriving market for 'A' licences (which fetch up to £1,000 despite the fact that there is no guarantee that they will be re-issued after the sale of a vehicle)[1] suggests that there is sufficient pressure to ensure an increase in capacity (and implicitly in cost) if the entry to the industry were freed. On the other hand the large amount of empty running currently undertaken by 'C' licence vehicles would be freed to take return loads with consequent decline in the demand for existing public

[1] To avoid this risk haulage businesses have changed hands in secret by the transfer of shares in limited liability companies without notification of the change of control to the Licensing Authority or of the changes in directors to the Companies Registrar. See the Report of the Licensing Authority for the Northern Traffic Area, 1961–2.

haulage. It is impossible to assess, *ex ante*, how these counter-acting forces would balance out, though we may hazard the opinion that the sum effect, in whichever direction prevailed, would not be great.

Secondly, there is the standard of service. It has been argued[1] that the increase in 'C' licence running in the last fifteen years indicates that public haulage has failed through its own inefficiency to achieve sufficiently high standards of service to compete with the carrier on own account. This argument is somewhat unfairly critical of the public haulier.

If we are to believe the T.R.T.A. survey, speed of delivery and certainty of timing are the main advantages which the 'C' licence operator has over the public haulier. This stems partly from the fact that his capacity is not as fully utilized. But, as the L.A.'s operate so that increased public road haulage capacity can only follow increased demand and not precede it, the public haulier is to a certain extent prevented from providing exactly those qualities of service which traders want. In these circumstances it is perhaps a little less than just to castigate the public haulier for his inefficiency; if the total capacity of the sector were unrestricted it might become more possible for the public haulier to provide services more competitive in quality with those of own account operation.

There is a danger, unfortunately, that improved service would only be available to the large consignor able to offer a substantial and regular demand, or for traffics on major routes. For less heavy and regular traffics the unpredictable interventions of the ancillary carrier might destroy the profitability of public haulage operation. The absence of a satisfactory network for the carriage of 'smalls' has already been specified as a major deficiency of the British transport system.[2] Paradoxically, in an unregulated system such traffic might only be disposed of with even greater difficulty and at greater cost. This might not be altogether a bad thing. We might argue that the high rates charged for such traffics truly represented the costs of carriage and that therefore it was ultimately in the public interest that such traffic should be subject to the restricting influence of high rates. On the other hand the resulting unreliability of 'smalls' service might have an undesirably inhibiting effect on business and the increased cost might induce further concentration of industry and population deemed undesirable on social grounds.

If we wished to liberalize the road haulage industry generally but

[1] P. Jenkins, 'Set the Haulier Free', *Crossbow*, New Year, 1962.

[2] See Sir R. Wilson, 'The Framework of Public Transport', *Journal of the Institute of Transport*, Vol. 25, No. 5, July 1953.

K

at the same time wanted to ensure the provision of a reliable smalls service at low cost, we might argue for the creation of a national light haulage monopoly, similar to the G.P.O., which could either use its own transport, or sub-contract, or even operate its own clearing house system to utilize empty return load capacity. A limited monopoly of this kind (of all consignments of less than 1 cwt not carried on own account, for example) might prove extremely difficult to enforce. On the other hand a smalls network operating a published tariff without such a monopoly position would inevitably incur losses as only traffic unprofitable at the declared rates would be left to it. Either way the provision for 'smalls' traffic is bound to be controversial.

Finally, to summarize the argument; in a fully employed economy providing adequate investment opportunities there is no reason to expect any disastrous expansion of capacity to follow the abolition of the present system of road haulage licensing. We might expect the ancillary haulier to gain some traffic from the public haulier to achieve better return loading; on the other hand the public haulier would probably compete more effectively for the provision of more refined services. In so far as the restrictions on road haulage were intended to protect the railways they have already failed due to the expansion of the 'C' licence capacity; in so far as they were intended to prevent cut-throat competition within the road haulage industry, changed economic conditions have made them redundant. Some liberalization would therefore seem appropriate.

TABLE 14

Numbers of Goods Vehicles analysed by type of body

	Number of vehicles				
	Licence category				
	Contract				
Type of body	A	A	B	C	Total
Tipper (excluding tanker)	7,400	8,500	29,300	61,500	106,700
Platform or sided	52,600	12,100	22,300	238,300	325,300
Box body	19,800	4,400	9,400	550,200	583,800
Tanker or other bulk carrier	1,600*	2,000*	1,300*	19,300	24,200
Livestock carrier	2,000*	—	2,300*	1,300*	5,600*
Other (including refrigerated van)	6,000	1,600	7,000	335,400	350,000
	89,300	28,600	71,700	1,206,000	1,395,600

* Sampling error probably large.

Source: Survey of Road Goods Transport, 1962. First Results.

TABLE 15

Numbers of Goods Vehicles analysed by carrying capacity

Number of vehicles
Licence category

Carrying capacity		Contract A	B	C	Total
Over	Not over	A				
—	1 ton	2,500	700	7,100	776,300	786,600
1 ton	2 tons	5,200	1,000	2,900	122,300	131,400
2 tons	3 tons	15,700	700	7,700	72,200	96,300
3 tons	5 tons	12,100	5,200	19,900	113,700	150,900
5 tons	7 tons	17,100	6,600	19,400	59,400	102,500
7 tons	10 tons	22,500	8,200	12,200	42,600	85,500
10 tons	13 tons	5,400	1,800	1,000	8,500	16,700
13 tons	—	8,800	4,200	1,500	11,100	25,600
Total		89,300	28,600	71,700	1,206,000	1,395,600

Source: Survey of Road Goods Transport, 1962. First Results.

TABLE 16

*'C' licence vehicles analysed by type of work on which the vehicle
was mainly engaged during survey week, April 2–8, 1962*

Percentages

	Unladen weight			
Type of work	Not over 2 tons	2 tons—3 tons	Over 3 tons	All vehicles
Retail delivery	34	31	16	31
Wholesale delivery	11	25	24	15
Maintenance and repair	15	3	3	12
Carriage of materials to or from building site	10	10	17	11
Delivery of materials and fuel to factories	2	5	12	4
Other	23	15	17	20
Not working	6	11	10	8
Total	100	100	100	100

Source: Survey of Road Goods Transport, 1962. First Results.

CHAPTER NINE

The Roads

Any comparison between road and rail transport is complicated by the fact that responsibility for the provision and maintenance of the track falls on the operator in the case of rail, but on the Government in the case of road transport. In view of the importance of road transport it is surprising that so little public concern has been devoted, until very recently, to the fulfilment of this public responsibility for the roads.

An inquiry by the Ministry of Transport in 1958 estimated that 56 per cent of all inland goods traffic, measured in terms of ton mileage, went by road[1] and in a similar calculation by the B.T.C. it was estimated that about 80 per cent of the passenger mileage travelled was undertaken on the roads.[2] Moreover, the pressure on the roads is growing. The number of vehicles on the roads tripled between 1946 and 1960 (from 3·1 million to 9·4 million), a cumulative growth rate of nearly 8 per cent per annum in the post war period. There is little to choose between the rates of increase of passenger and goods vehicles but in both cases there has also been an increase in the average size and speed of vehicles. During this same period the road mileage in Great Britain only rose from 183,000 to 196,000. Hence a 200 per cent increase in traffic has been met by about 7 per cent increase in available road mileage. The number of vehicles per road mile consequently increased from seventeen in 1946 to forty-eight in 1960. Though these figures tend to over-emphasize the disparity between the demand and the supply of road space,[3] there can be little doubt about the neglect of the road system in the early post war years in particular. It is not until the last five years that there has been any really significant expenditure on new construction and major improvement.

If the money expenditures are deflated to take account of the rise in prices some extremely instructive comparisons are obtained. For instance, total expenditure on roads was less in real terms in 1949 than it had been in 1911 despite the fact that since that time the

[1] *The Transport of Goods by Road*, H.M.S.O., 1959, p. 27.
[2] Report of the Select Committee on Nationalized Industries, July 1960, Appendix 8, p. 335.
[3] Because they do not take into account the relief produced by road widening.

TABLE 17

Spending on Public Roads

£000,000

	1939	1947	1951	1956	1960	1962
Total comprised of:	58·3	60·6	72·8	111·1	188·0	223·3
1 New construction and major improvement	17·4	2·5	6·3	13·1	72·6	91·4
2 Maintenance and minor improvement	31·6	41·6	49·1	73·8	84·7	97·3
3 Cleansing, administration, etc.*	9·3	16·4	17·4	24·2	30·7	34·4

* After 1951, cleansing on trunk roads transferred from 2 to 3.
Source: Annual Abstract of Statistics, 1962.

number of vehicles in use had increased more than twenty-fold.[1]

This disappointing performance in immediate post war years does not reflect the intentions of the Labour Government but its reaction to crisis. A ten year plan for road development was announced in 1946 which included a planned expenditure of £80 million per annum during the first two years. Unfortunately the cuts in public investment subsequently forced by crisis conditions bore very heavily on the roads plan. As a result, expenditure on new construction or major improvement only amounted to a total of £37 million for the seven years between 1947 and 1953, an average of just over £5 million per annum. If we were to make allowance for depreciation during this period, real *net* fixed capital formation in roads would have been negative.

At last, in 1953, the roads began to attract attention and in December of that year a three year, £50 million, investment plan was announced. This was superseded before its completion by a plan, extending to 1959, authorizing total expenditure of over £200 million. Again in July 1957 it was announced that schemes costing £280 million would be authorized in the period 1958–9 to 1961–2. Perhaps even more significant than the size of this plan was its exclusion from the general cuts in investment made in 1957. Of similar significance was the announcement, in 1960, that road investment was to be further increased by some £20 million per annum whilst rail investment was to be decreased by the same amount. Clearly the roads programme had at last achieved high priority.[2]

[1] See P.E.P., 'The Cost of Roads', Table 2, p. 115. (*Planning*, No. 452.)

[2] It has been estimated that a total investment of over £3,000 million would be required during the decade 1960–70 to make reasonable provision for traffic, assuming it to grow at the same rate as in the last decade. See D. J. Reynolds, 'The Outlook for Expenditure on Roads', *London and Cambridge Economic Bulletin*, No. 37, March 1961. The planned expenditure for 1964–9 is £865 million. See *Roads in England and Wales*, 1963, p. 28.

HIGHWAY ADMINISTRATION

Now let us examine briefly the way in which these funds are disbursed. The present structure of highway administration dates back in principle to the Roads Act of 1920 which provided for the Ministry of Transport to make grants to local authorities towards the cost of highway maintenance out of a 'Road Fund' into which all receipts from licence duties and fuel taxation were paid. A classification of roads was undertaken, the extent of Ministry aid being graduated according to the importance of the road as a main routeway. Since then the classification has been expanded and the proportion of costs met by Ministry grants has been increased. In 1956 the Road Fund was abolished and the responsibility for finance transferred directly to the Ministry of Transport in England and Wales and to the Scottish Home Department in Scotland. The general principles of highway administration remained unaltered however. The composition of the British road system in 1961 and the aid granted was as follows:

TABLE 18

The British Road Network, Spring 1961

Class	Mileage	Grant
Trunk and Motorway	8,468	Ministry 100% responsible
Class I	19,747	Ministry 75% responsible
Class II	17,620	Ministry 60% responsible
Class III	48,927	Ministry 50% responsible
Unclassified	100,455	All expenditure rates financed

Sources: Roads in England and Wales, 1961–2. Scottish Roads Report, 1961.

Despite the predominance of unclassified roads in mileage terms, the bulk of the new investment will be directed towards trunk and Class I roads. This is partly because it is these roads which are congested and therefore deemed to require improvement and partly because, despite increased grants to local authorities for the relief of urban congestion, the chief priority has been given to the five major projects which comprise the main arteries for industrial road transport.

THE PRINCIPLES OF ROAD EXPENDITURE

The high priority now given to road investment focuses attention on the absence hitherto of any satisfactorily objective criterion for assessing the desirability of road projects. Even if desirable projects are far in excess of the funds currently available such a criterion is still necessary to establish a scale of priorities for the allocation of the limited budget. We must consider then the various principles on which investment in roads might be based.

A first possibility is suggested by orthodox Keynesian macro-economic theory. Public works in general, and road works in particular, are amongst the prescribed Keynesian antidotes to general unemployment. Consequently we would expect to find road expenditures high in times of depression and low in inflationary periods. Unfortunately constant inflationary pressure in the immediate post war years led to the almost total exclusion of investment in roads. Hence, if the road system of the United Kingdom was to be improved, this approach to road investment had to be rejected, for other methods of control over the economy had been developed sufficiently to preclude the possibility of a depression long enough, sufficiently severe or appropriately concentrated geographically to enable the leeway in road investment to be made up. It was symptomatic of such a change in attitude that road expenditures were specifically excluded from the cuts in public investment made in July 1957.

In contrast it has been argued that as receipts from the taxation of road use usually exceed the annual expenditure on the road system then there are grounds for either a reduction in taxation or an increase in road investment. The facts are indisputable; in 1961, for example, expenditure was only about 45 per cent of receipts (see Table 19) even if we include all expenditure on roads and public lighting together with half the costs of the police force as attributable to road users.

TABLE 19

Road Taxation Receipts and Road Expenditure, 1961

Receipts (£m.)		*Expenditure (£m.)*	
Motor vehicles		Roads and Public	
licence duties (a)	141	Lighting (c)	283
Hydrocarbon oil		Police force (50%	
tax (a)	480	of total) (c)	80
Purchase tax on			
vehicles (b)	166		
Total	787		363

Sources: (a) National Income and Expenditure, 1962, C.S.O. Table 36.
(b) Report of Commissioners of H.M. Customs and Excise, p. 103.
(c) N.I.E., 1962, Tables 43 and 44.

Nevertheless it would be wrong to use this as the basis of an argument for increased investment in roads for neither direct nor indirect taxation is normally linked to the particular benefits which the taxpayer receives from the administration. Smokers, for instance, contributed over £800 million to the Exchequer in 1961–2 in return for which the only expenditure directed specifically to them was an ungrateful publicity campaign decrying smoking as a risk to health.

Neither of the methods of determining the amount to be invested in roads which we have so far considered is at all satisfactory. In the first case there is a tendency for investment to be retarded and beneficial opportunities wasted; in the second case a danger of investment being undertaken where no real benefits exist, with a consequent waste of resources. In order to avoid both of these contrasting perils we need to devise a method of comparing the benefits ensuing from any given investment with the cost of the investment. In most sectors the price mechanism and the profit motive is accepted as a rough, but generally satisfactory, means of securing that resources are only put to uses which the consumer values more highly than the alternatives foregone. Why then cannot we treat roads in exactly the same way?

Prices for roads
Though, with the exception of a few very limited stretches of toll road, there is no direct pricing system operated on our roads, in a sense we can look at licence fees and fuel taxation as the price of motoring. The licence fee is an overhead charge, the effect on cost per mile varying inversely with the mileage covered. The petrol duty varies directly with mileage, in a way determined by the rate of petrol consumption of the vehicle involved. The marginal payment for use of the road paid in this way by private cars is roughly between ¾d and 1¼d per mile.[1] On the basis of published statistics for maintenance expenditure on roads and the Road Research Laboratory censuses of traffic flows, Mr D. L. Munby has made estimates of average maintenance cost per vehicle mile on various classes of road. These range from ·1d per vehicle mile on trunk and Class I roads in Scotland to 1·1d per vehicle mile on Class II roads in rural England. On almost all roads except the Class II and III roads in rural England and Wales the maintenance costs per vehicle mile are less than the price paid per vehicle mile at the margin in the form of petrol duty. The reason for the high cost per vehicle mile in the exceptional cases is the low level of utilization of the roads. Even so, the marginal revenue derived by the Government from motorists seems to be generally greater than the average cost of road maintenance. As

[1] See D. L. Munby, 'The Roads as Economic Assets', *Bulletin of the Oxford Institute of Statistics*, Vol. 22, No. 4, November 1960.

maintenance is in any case more a function of time than a user cost the argument holds *a fortiori* for short run marginal cost. Hence under the present dispensation price is not related to cost in any sensible way.

Now let us look at how the price mechanism might be used as a guide to investment in a new road. There are two distinct functions that we normally expect a price mechanism to fulfil:

(i) It must suggest what new facilities to provide.
(ii) It must tell us how existing facilities are to be used.

In using the price mechanism to these ends, there are two crucial questions to be answered:

(i) What costs should in total be covered?
(ii) How should costs be covered?

To begin with we would argue that the normal commercial criterion requires that investment takes place if receipts from a new asset would be sufficient to cover total costs, including the appropriate costs of capital. The total annual cost for a road consists of the following items:

(a) Charges on capital, at the relevant rate.[1]
(b) Depreciation (though it may be over a long period).
(c) Maintenance costs.
(d) Lighting.
(e) Signalling, police, etc.
(f) Administration.

If these total costs can be met then on normal commercial grounds the project would seem to be acceptable.

Unfortunately this will not necessarily stand by itself as a criteria similar to that normally applied in business investment. No firm could hope to make a profitable investment if its competitors were giving away similar products without charge. There would just be no demand. Similarly we are only able to judge the new investment if alternative facilities are being provided at a charge not less than the real cost of their provision. Hence we must institute a scheme which charges not only for new roads but also for existing roads if the price mechanism is to be used as a guide to investment.

In deciding what are the relevant costs to be charged to users of existing roads, let us begin with those roads which have plenty of

[1] For a discussion of the relevant rate see C. D. Foster, 'The Cost of Financing the Nationalized Industries', *Bulletin of the Oxford Institute of Statistics*, Vol. 22, No. 2, May 1960. Also C. D. Foster, *The Transport Problem*.

spare capacity, that is they are not congested,[1] and there is no question of any replacement or improvement being necessary. In this case the real cost to the community is likely to be relatively low, comprising only wear and tear due to use, which ought to be charged directly to the individual user if optimum use is to be encouraged at the margin, and the maintenance and signalling costs required to keep the road open, which should be recovered from users collectively if the road is to be maintained. As we are talking about the 'real' or 'opportunity' cost to the community of the road, its historical cost is irrelevant. The only other consideration which would appear to be relevant is the possibility that the land could be put to alternative uses. G. J. Ponsonby has argued in reference to railway tracks[2] that the earnings which could be obtained in an alternative use are true cost to the community and should be covered if the assets are to remain in their current use. It is unlikely, however, that this will be a very important element in the case of uncongested roads for the costs of transfer into farming land are high and land values are often dependent on the existence of road facilities and would fall if the transfer destroyed these facilities. The 'real' costs of use of uncongested roads then may be relatively low; it is nevertheless important to charge them to the users of the roads because of the income effect that this will have on their demand for the rest of the road system.

In the case of a congested road the concept of 'opportunity cost' is much more complicated. Each vehicle using the road adds to the congestion and so from the community's point of view the real cost is not only the maintenance, signalling and policing cost of the road but also the extra cost which is incurred by all other road users as the result of the additional vehicle. These would consist of the extra fuel used, the value of the time wasted by the delay and also possibly some cost of the sheer frustration and annoyance caused!

These extra costs resulting from a road usage decision may be referred to as the marginal social cost. A rational decision on road use would seem to require that no individual be permitted to undertake any action unless he valued it more highly than the marginal social cost. Therefore the price for each individual road user should be at least equal to the marginal social cost.

A simple example will demonstrate the usefulness of a toll system

[1] We shall consider congestion to commence to exist where the addition of one more vehicle reduces the speed of transit of those currently on the road.
[2] G. J. Ponsonby, 'Earnings on Railway Capital', *Economic Journal*, Vol. LXX, No. 2, June 1960.

in solving the problem of optimum use of existing facilities.[1] Let us consider alternative roads between two places. The amount of traffic offering for the journey is fixed at amount AB in Figure 10. We may assume with some realism that as more traffic uses a road, congestion, and hence the time and cost of the journey, increases. But the rate at which cost varies with respect to increasing traffic will depend on the physical characteristics of the road.

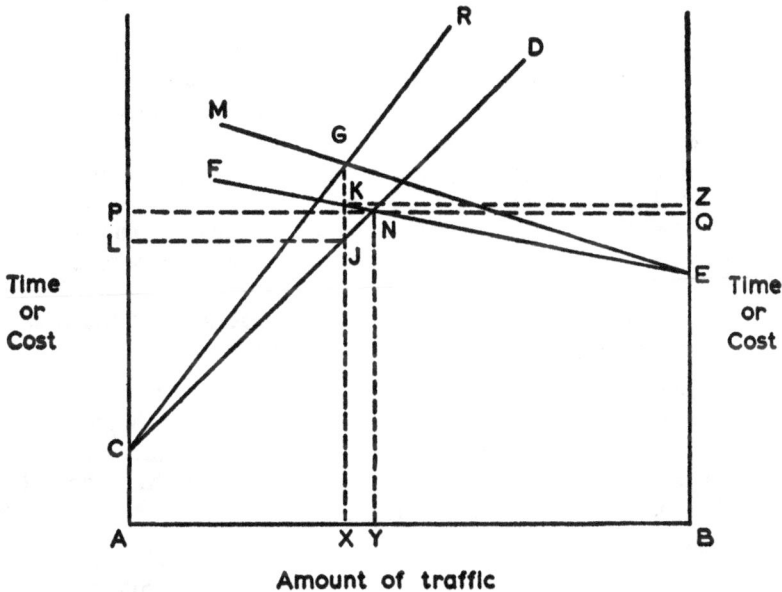

Figure 10

In our diagram differing characteristics are shown by CD (a fast road when empty but of low capacity) and EF (a slower road with greater capacity). Now, if we assume that each driver is interested only in his own cost, and that this is directly proportional to the time taken, he will use whichever road promises the shortest journey time. If all drivers operate in this way and there is perfect knowledge of the current state of congestion on the alternative roads traffic will be distributed so as to equalize transit time on the two roads. But, total cost could be reduced if some traffic, which under free choice will travel on road A, could be persuaded to travel on

[1] Based on an example in Beckmann, McGuire and Winsten, *Studies in the Economics of Transportation*, Yale, 1955, p. 84, the argument was first expounded by Professor Pigou in *The Economics of Welfare*, London, 1920, p. 162.

road B. This is because the sum of the extra cost incurred by the vehicles transferred plus the increase in cost for all other vehicles using road B due to extra congestion caused by the transfer is less than the total reduction of cost accruing to those units still running on road A as a result of reduced congestion. The rates of change of total cost as extra traffic is taken on the two roads are shown by the curves CR and EM. These are the marginal cost curves and the total cost for all the traffic AB will be minimized when these marginal rates for the two roads are equalized. This point is shown at G in our diagram. The total cost for traffic AB is now equal to ALJX plus BZKX and this is less than the total cost APQB which would be achieved if the allocation of traffic between the two roads was made by free choice. Whereas free choice takes no account of marginal social cost, i.e. the cost incurred by other road users as the result of an individual's decision, a toll on road A equal to the current difference between marginal social cost on the two roads would ensure that the total real cost of transport between our two locations was minimized.

This would suggest that in the case of a congested road the price charged should include two elements, namely:

(i) A marginal rate reflecting marginal social cost.
(ii) An overhead charge for permission to use the facility or group of facilities sufficient to meet all costs not directly variable with use, raised in such a way (i.e. on the principle of 'what the traffic will bear') as to leave the marginal decisions unaffected.

The generation of a surplus of receipts over expenditure by part (i) of this charge would indicate the consideration of constructing additional facilities. If the annual surplus on the existing road, which would be eliminated if a new road was built, was sufficient to meet the total annual cost of a new road then consumers would have indicated that they valued the total new service more highly than its cost and hence it should be implemented. If capital were limited then it would be preferable instead to implement only those schemes which would have greatest effect per £ invested in eliminating surpluses.

This system would, incidentally, automatically involve the allocation of land between road and other uses as a high price for land would reduce the likelihood of a scheme being implemented and vice versa.

This still leaves us with the problem of deciding what value to attribute to marginal social cost in any instance. In principle we would need to measure the effect of adding an individual vehicle to the road at the time under consideration. In physical terms this would cause wastes of motor fuel, time and nervous energy. The first of these might be easily evaluated. Attempts to value time losses have been made[1] but are, of necessity, rough and ready and involve heroic assumptions. Perhaps the least objectionable method of evaluating these savings, if it were possible, would be at user's valuation; that is by finding out exactly how much each individual involved would pay to avoid the congestion caused. If we were able to discriminate perfectly to take from each consumer his whole consumer surplus any losses to road users would be reflected in revenue losses, thus automatically giving a valuation. We could then simply require the road authority to price in such a way as to maximize its surplus in order to obtain maximum benefit. In the absence of such perfect discrimination marginal social cost remains a very difficult quantity to evaluate.

Nevertheless, in principle there is much to be said for a pricing system for roads, both as a guide to use of existing facilities and to investment. If some practicable way could be found of implementing the method it would certainly improve on present methods in both respects.

Before going on to consider the practicability of pricing for roads we ought to mention three defects in principle of the method which we shall ignore:

(i) Discriminatory pricing in transport, implied by our method, may create a bias in favour of investment in and use of assets in this sector.

(ii) In contrast, the fact that discrimination is imperfectly practised means that within the transport sector there is not necessarily a distribution of resources perfectly in accordance with consumers' valuation.

(iii) No allowance has been made for the consideration of social benefit or cost in the wider context of the aesthetics of various alternatives.

The practicability of pricing for road use

The reason why such a pricing scheme is not *at present* in operation is one of practicability. A contributory factor in the decline of the Turnpike Trusts in the nineteenth century was the unpopularity and sheer inconvenience of toll gates. There are now very few facilities for which tolls are charged and in those cases that remain the time spent paying the toll is insignificant in comparison to the time saved

[1] See pp. 160-2.

by the use of the particular route. The toll is thus unlikely to cause any serious diversion of traffic away from the route. If, however, we were to attempt to use tolls as a method of pricing roads generally, the practical difficulties would be immense. The road network is so intense that in most cases there are alternative routes which would all have to be tolled, at different rates, if uneconomic diversion of traffic was to be prevented from toll on to free roads.

The use of tolls for congested *areas* as opposed to individual roads might help solve the problem of congestion in town centres whilst not affecting the actual use of the centre by people as opposed to cars. The same effect might be achieved if it was necessary to hold a special licence for use of particular areas, similar in a way to the tickets carried by London taxis.[1] In this case there might be an even greater reduction in traffic as people attempted to avoid incurring the original overhead cost of the special area licence.

Recently it has been suggested that the toll and the licence are both archaic ways of charging for the use of roads; if we really wished to charge economic prices with greater precision the mechanics of the system of collection could certainly be solved in a more scientific manner.[2] Two different types of mechanical charging device have been suggested so far.

(a) *Meters on vehicles*. There are two variations of this method. R. M. Walsh has devised a meter incorporating an electrolytic timer which shows a light when it is in operation. This system could be made fairly sophisticated by having several different rates of consumption and automatic switching without the cost of the meter exceeding £5 per car. This is a time charge. A second variation would use a meter in the car simply to record impulses directed at it from the road surface. Again the method should not prove unduly costly and could be made as precise (by varying the density of impulses) as required. Both methods could be used to charge automatically for parking in congested areas. A pricing system on these lines was first proposed by General Precision Systems Ltd. of Aylesbury. These systems are akin to the methods of charging for gas and electricity with which we are familiar.

(b) *Scanning systems*. In this system, first suggested by Professor William Vickrey as an answer to the problem in Washington D.C.,

---

[1] In Malta, for instance, a special tax of £3 per annum is charged to private cars entering Valletta.
[2] See G. J. Roth and J. M. Thompson, 'Road Pricing—a Cure for Congestion?', *Aspect*, April 1963.

vehicles would carry electronic identification plates which would be scanned and recorded at various pricing points throughout the city. The system could be made completely automatic, right down to the preparation of quarterly accounts, by the use of computers. It has this advantage of being more automatic than others, but on the other hand it would be extremely costly to install. For instance Pye Telecommunications Ltd. have estimated that the cost of the equipment for a small town like Cambridge would be about £400,000.

These suggestions yield problems for the electronic engineers. The economist is only directly concerned in the solution in so far as he must weigh the costs of introducing the necessary charging mechanisms against the advantages to be derived from their operation. These advantages may be substantial for a metering system able to register differential charges according to the time of day or state of road congestion thus ensuring a constant incentive to avoid use of the roads at peak periods. As yet, however, it has not been conclusively demonstrated that a pricing system is practicable.

Investment decisions must be taken, however, even if we are not able to operate a pricing system. We must therefore devise an investment criterion which is not dependent on the existence of an economic price scheme. Ideally we might argue that, assuming the amount of available investment funds to be limited, we ought to implement those schemes which achieve the greatest increases in National Income per £ invested in whatever sector they arise. Unfortunately it is virtually impossible to fully identify those changes of National Income caused by a new investment. Nevertheless, if we could trace and value all the benefits arising from investments we might be able to compare the value for money of different schemes, i.e. the rate of benefit per £ invested. This is the 'social surplus criterion'.

Cost-benefit analysis: the social surplus criterion

The prototype of this form of analysis was the investigation of the economics of the London–Birmingham motorway, the M1.[1] This investigation was undertaken after the decision to build the road had already been reached; fortunately it concluded in favour of the project, though it did suggest that alternative projects may well have been preferable. Since this first major investigation the approach has been generalized to deal with comparison between alternative road

[1] T. M. Coburn, M. E. Beesley and D. J. Reynolds, *The London–Birmingham Motorway*, Road Research Technical Paper No. 46, 1960.

investment projects[1] and it has also been applied to one railway project, the Victoria line.[2]

We shall be concerned only with the principles on which the economic assessment is based, set out in the second part of the published paper on the M1. It was assumed that the price of alternative forms of transport would remain stable. Inquiries were carried out on alternative roads, the A5 and A6, to determine how much traffic might be expected to transfer to the new road on the assumption that traffic would choose whichever route would minimize journey time. Three alternative sets of assumptions were made concerning the speeds achievable on the M1, giving three alternative assignments of traffic. The real savings expected to result from the construction of the new road were then estimated for each set of assumptions. The savings may be classified as follows:

(a) Savings by traffic using the M1.
   (i) Vehicle costs:
      (a) Labour.
      (b) Capital.
      (c) Fuel.
      (d) Other operating costs.
   (ii) Accident costs.
(b) Savings by traffic remaining on other routes by virtue of reduced congestion.
(c) Other social benefits.

These savings were then transformed into a common monetary standard. For working hours saved an estimate was made of the relevant hourly wage rate for the purposes of evaluating the savings, though it was not, in the first instance, deemed possible to make any such well-founded evaluation of leisure time saved. In the case of vehicle stock it was assumed that the savings were directly proportional to the vehicle hours saved (i.e. that time savings could be effectively utilized) and that capital freed in this way could yield 15 per cent per annum in other uses, in order to transform them into financial savings. Estimates were similarly made of the value of fuel savings and of other vehicle savings such as tyre, brake and clutch wear on the basis of assumed vehicle performance. Finally, in assessing the saving in accident costs some evaluation had to be

[1] D. J. Reynolds, *The Assessment of Priority for Road Improvements*, Road Research Technical Paper No. 48, 1960.
[2] C. D. Foster and M. E. Beesley, 'Estimating the Social Benefit of Constructing an Underground Railway in London', *Journal of the Royal Statistical Society*, Series A, Vol. 126, Part I, 1963.

made of human injury as well as other losses through accident. The estimated annual saving in this category was stated with some hesitance.

Against these total savings two additions to cost had to be offset. Firstly, as the assignment of traffic to the M1 was based on the time element alone there would inevitably be some traffic which, though saving time by using the faster route, would incur extra fuel costs by covering a greater mileage. This item was evaluated separately for the three different traffic assignments. Secondly, the extra annual maintenance cost of the road system resulting from the use of the M1 was also set off against the benefits.

The net savings, or benefits, ensuing from the project were then compared with the capital costs of the project and expressed as a rate of return on capital. The estimated total capital costs of £22·46 million was inflated to £23·3 million to allow for interest charges at 5 per cent during the period of construction. The resulting table of returns is shown as Table 20.

TABLE 20

Social Benefits Resulting from the Construction of the M1

Changes in Costs (£000's per annum)

	1st assumptions	2nd assumptions	3rd assumptions
Savings in working time	−453	−624	−766
Reduction in vehicle fleets	− 80	−161	−227
Reduction in fuel consumption	−117	− 84	− 18
Other operating costs	−200	−200	−200
Costs of added mileage incurred	+229	+307	+375
Reduction in costs of old roads	−128	−128	−128
Total vehicle costs	−749	−890	−964
Reduction in accident costs	−215	−215	−215
Maintenance costs of M1	+200	+200	+200
Net annual savings	−764	−905	−979
Expressed as return on capital	3·8%	4·5%	4·8%

Source: Beesley, Coburn and Reynolds, *op. cit.*

These returns might be inflated on three grounds.

Firstly, there is no allowance for the generation of entirely new traffic by the building of the motorway. It was argued that in this case little could expect to be transferred from the railways and that the benefits accruing to any marginal traffic brought forward would, by the very fact that it was marginal traffic, be relatively small.

Secondly, the assessments made do not allow for any value to be

L

placed on non-business time. The rates of return would vary according to the valuation of leisure time (see Table 21).

TABLE 21

M1 Sensity of Rate of Return to Valuation of Leisure Time Savings

Average values of leisure time saved	Rates of return %		
	1st assessment	2nd assessment	3rd assessment
0	3·8	4·5	4·8
2s per hour	4·6	5·4	5·9
4s per hour	5·4	6·4	7·1
6s per hour	6·2	7·4	8·3
8s per hour	7·0	8·3	9·4
10s per hour	7·8	9·3	10·5

Source: Beesley, Coburn and Reynolds, *op. cit.*

Thirdly, the assumptions used are those of zero growth of traffic. If, however, projections were made into the future of recent rates of growth of traffic then the rates of return would be considerably higher (see Table 22).

TABLE 22

Expected Future Rates of Return on M1 Investment

Present (1958)	4·6—10·5% (minimum and maximum values from Table 2)
1960	9·9—15·2% (minimum and maximum values from Table 2)
1965	17·6—27·3% (minimum and maximum values from Table 2)

Source: Beesley, Coburn and Reynolds, *op. cit.*

The meaning of the analysis

In the interpretation of their results the authors say: 'In order for the construction of the motorway to be worthwhile it must be shown that the rate of return obtainable is greater than the current rate of interest, and more rigorously, greater than the rates of return obtainable in other uses of capital, including the other road improvements.' As the relevant rate (yield on Consols) was about 5 per cent during the period of the investment, then the project would seem to meet the first requirement. Unless all other capital projects were similarly examined however, there could be no assurance that the project would meet the second requirement.

The authors themselves admitted four possible sources of error in their calculations, namely:

(1) There is no satisfactory valuation of leisure time saved.
(2) Factors other than time savings may effect the transfer of traffic to the M1.

(3) Speeds may be increased on existing roads causing savings.
(4) The incidence of fuel taxation introduces distortion which causes too much emphasis to be placed on fuel savings and not enough on time savings.

To these we should add the criticisms that have been made since the publication of the paper.[1]

Firstly it has been argued that this type of assessment may give a misleading indication of where new investment is desirable if present resources are being inefficiently used. One of the prerequisites for an economic use of resources is that price should be similarly related to cost in competing sectors. The absence of this condition in the case of roads does not necessarily invalidate the method as an expression of the benefits expected to be realized from a particular scheme under current conditions. It does not tell us, however, whether the same benefits could be achieved by a more simple structural change.

Secondly, we may question the valuation of time savings at hourly wage rates, which may not be the relevant rates for two reasons:

(a) Savings of working time may be taken in the form of increased leisure. As the hourly wage rate includes payment for time and for effort, then the value of the leisure (i.e. the value of the time element only), which is relevant here, will be less than the total hourly wage rate.
(b) In any case the savings in real terms should be the savings involved in after tax earnings, not in before tax earnings, as for the community as a whole the loss of Government revenue must be offset against the entrepreneur's savings in wage costs.

Finally, we come to the crucial question of comparability. The authors state at the beginning of their economic assessment that 'The rate of return (calculated as we have shown) may then be used for comparison with other forms of investment and with the rate of interest to decide whether the investment is worth-while.' If by this they are referring to the rates of return in other spheres *similarly calculated by valuation of surpluses* we may take their point, but the reference to the rate of interest in this context suggests that they were referring to the commercial rate of return on other investments. Such a contention would not, given this interpretation, be acceptable. We can perhaps emphasize this point by reference to the two schemes which have been discussed recently.

[1] See D. W. Glassborow, 'The Road Research Laboratory's Investment Criterion Examined', *Bulletin of the Oxford Institute of Statistics*, Vol. 22, No. 4, November 1960.

In the case of the Victoria line addition to London's underground railway it was estimated that there would be a total financial loss of £3 million a year after payment of interest on the capital invested. Hence this would be a bad investment on normal commercial grounds. However, a cost-benefit analysis, similar to that carried out for the M1, yielded the conclusion that great benefits would accrue both to road users and to users of other underground services, which, added to the benefits obtained by users of the new line itself would appear to constitute a good return on the capital involved.[1] Clearly it is this return which is comparable to the rate calculated on the M1.

Another example is the Channel Tunnel scheme. Even on the commercial test this is a project which may be profitable, but because of the size of the project and its political implications it would probably require some finance from Government sources. Now the social return of the scheme, calculated on surplus grounds, would undoubtedly be considerably higher than the strictly commercial return. Thus it would be clearly wrong to reject the scheme on the grounds that the expected commercial return on it is less than the social return forecast for some of the alternative schemes of road improvement. Thus we cannot justifiably compare rates of return on capital investment in transport unless the rates of return are similarly calculated.

Even if this technique cannot be used for comparison of the rate of return on road schemes with the commercial returns on projects in other spheres there is no reason why, in the absence of any better suggestion, it should not be used for comparison of alternative road projects. An earlier application of cost-benefit analysis to the problem might have significantly altered the whole pattern of road investment in the last five years. The Ministry of Transport, in evidence given to the Select Committee on Estimates in 1959, claimed that it concentrated on those schemes which would most benefit industrial traffic. The Select Committee was not entirely convinced. It criticized the Ministry for deciding the priorities on the grounds of cost alone, without relation to the benefits arising, and for its willingness to permit national priorities to be determined to a certain extent by the enthusiasm of local authorities. The Committee recommended that higher priority be given to urban improvements. Pilot studies at the Road Research Laboratory supported the recommendation by showing that higher rates of return could be obtained by road widening, road straightening and the improvement of intersections in urban areas than by the major rural schemes.[2] It has even been

[1] See C. D. Foster and M. E. Beesley, *op. cit.*
[2] D. J. Reynolds, *op. cit.*

argued that a 2×2 lane M1 would have been a better investment than the 2×3 lane highway actually constructed.[1]

CONGESTION IN TOWNS

There are some characteristics of the roads problem which are specific to urban areas.[2]

We can certainly demonstrate, in physical terms, the superiority of public as opposed to private transport in the use of urban road space. Dr R. J. Smeed has estimated[3] that in the peak period three private cars require as much road space as a bus. Taking average loadings, $4\frac{1}{2}$ people carried by private transport thus require as much space as 27 carried by public transport. Moreover, the greater the pressure on road space the less favourable is any such comparison to private transport.

The cost of urban road improvement and construction is immense, reaching more than £10 million per mile in some cases. Though immense benefits are necessary to justify such investment, the greater cost does not involve any difference in principle. The urban road problem is bedevilled by one important complication however. It was assumed, in the M1 assessment, that there would not be any significant generation of new traffic. This assumption may not hold for urban improvements. New traffic may be generated either by an increase of the number of journeys made or by a transfer from public to private transport. In some cases traffic may be so responsive to road conditions in this way that the expected benefits to existing traffic would be almost totally eliminated. Where improvements generate so much additional traffic that congestion is not eliminated, any assessment simply in terms of existing traffic might seriously over-estimate the benefits obtainable. This effect could only be incorporated in the cost-benefit analysis by including an estimate of the potential traffic response to road improvements.

Parking
Before incurring the tremendous cost of construction of roads it is sensible to check on available alternatives. In many towns street parking is a major cause of congestion. For example, it has been

[1] M. E. Beesley, 'Some Aspects of the Economics of the M1', *Journal of Industrial Economics*, Vol. 10, No. 3, July 1962.

[2] For a much fuller discussion of the urban road problem see Alan Day, *Roads*, London, 1963.

[3] R. J. Smeed, 'The Traffic Problem in Towns', Manchester Statistical Society, February 1961.

estimated[1] that the removal of 100 parked vehicles per mile of road in the London area would have the same effect on traffic flow as a widening of 5 feet. Provision of off-street parking might, therefore, achieve a given increase in traffic flow more cheaply than road widening. Realizing this, some towns, such as Coventry and Bedford, combine restriction of street parking with provision of adequate off-street parks in the central area to very good purpose. It is even found that the construction of multi-storey car parks can be a feasible commercial proposition.

Unfortunately the provision of central parking space may simply generate new traffic which will cause other bottlenecks to appear.[2] The real cost of parking facilities in a case like this would be not only the actual land and construction costs of the park itself but also the extra costs involved in reducing the generated congestion in other parts of the town. Thus, under the present system whereby the parking charge is the only toll levied on town motoring, it does not follow that improved central parking facilities are socially justified simply because they can be operated as commercial propositions.

SUMMARY

Following a long period during which investment in roads was negligible, considerable pressure has recently been applied to obtain improved road facilities. Unfortunately, because of the absence of a price system for roads it is very difficult to assess what is the 'effective demand' for new roads. The introduction of some form of price mechanism and the use of surplus criteria have both been suggested as objective methods of assessing the social desirability of roads. Neither is without flaws, both of principle and of practicability, but at least their consideration has caused some light to be shed on the real costs of road use. For example, some types of transport are more 'space-expensive' than others. It seems quite clear that if vehicles were charged differentially according to the scarcity value of the road space that they consume, there would be more incentive than exists at present for the use of public transport in urban areas. Therefore, unless we are willing to incur the much higher costs involved we must beware of encouraging (or failing to discourage) too much transfer from public to private transport in urban areas. At the same

[1] G. Charlesworth and J. L. Paisley, 'The Economic Assessment of Returns from Road Works', Institution of Civil Engineers, November 1959.

[2] Such a vicious circle is suggested by American experience, see P. E. P., 'Solving Traffic Problems—Lessons from America', *Planning*, No. 402, September 1956.

time, however, there is no justification for preventing the spread of the greater flexibility of private transport so long as it does not involve social costs over and above the private costs or expenses of motoring.

In conclusion, we may add a word of warning. A piecemeal approach to the problem of congestion by means of the successive elimination of the worst points may only involve us in a vicious circle of congestion, construction and regenerated congestion unless it is also linked with the efficient use of existing assets resulting from some relation between costs and charges. In this respect the report of the Buchanan Committee, published in November 1963, is disappointing as it omits consideration of a revised road charges scheme.[1] The accompanying Crowther group report considers the possibility briefly but dismisses it on the grounds that to be effective charges would need to be heavy, and use of urban roads would therefore be limited to the rich. Such a curt dismissal does not do justice to the merits of the case. The value judgment that urban road space should, like medical treatment, be available to the individual on demand irrespective of its cost to the community might not appear so acceptable if the real costs of such a course and the alternatives possible were more thoroughly investigated.

[1] *Traffic in Towns.* H.M.S.O., 1963.

CHAPTER TEN

The Railways

Neither the Government nor the public appear to be in very much doubt that the railway system currently poses a very serious economic problem. The financial record of the B.T.C., which has been consistently in deficit since 1953, is taken as sufficient evidence of the magnitude of the problem. Moreover, if we distinguish between the railways and other activities the railway record shows up in an even worse light (see Figure 11).

B.T.C. FIANANCIAL RESULTS 1948-1962

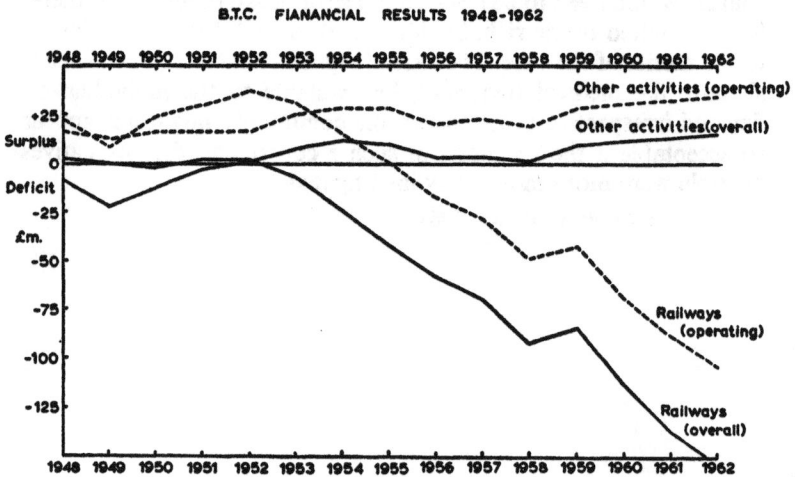

Figure 11

Note: The Central Charges, which account for the difference between the operating result and the overall figure, consist predominantly of interest on capital but also include the cost of central administration and the redemption provisions. Interest on accumulated losses since 1958 and on new borrowings 1957–60 transferred to special account is not included at all. By 1960 this amounted to another £30 million. Central charges were not allocated between railways and other activities in the B.T.C. accounts before 1956. For the purpose of our diagram however they have been allocated in the same proportions as in 1956.

On working account the railways continued to show a surplus until 1956 but only in one year, 1952, was this surplus large enough to meet the railway share of central charges. Though not all of the other activities have been profitable (British Waterways, for instance, regularly incurred losses), in total a modest profit has always been shown by the other activities. The B.T.C. loss appears then to be primarily a railway problem.

Many explanations and interpretations of this situation have been produced. We shall begin our investigation of railway economics by examining some of these arguments.

EXPLANATIONS OF RAILWAY DEFICITS

(i) Over-extended capital structure

It has been argued that compensation payments for railways assets taken over in 1948 were unduly generous.[1] The Stock Exchange valuation at the end of the war, on which the terms of compensation were based, reflected a degree of optimism resulting from the large, though forced, transfer of traffic from road to rail during the war. Moreover, the railway shares were valued in relation to a $2\frac{1}{2}$ per cent gilt edge rate but the Transport stock eventually issued at 3 per cent. As a result the annual interest payments required of the B.T.C. were unrealistically high in relation to the real value of the assets acquired. In addition, the obligation to redeem all British Transport stock over a period of 90 years added a further fixed annual charge which would not have been required of a private company.

The excessive size of the commencing debt, though a contributary factor, is far from being a sufficient explanation of the financial shortcomings of the railways. If we exclude central charges completely (and this is to exclude not only interest payments but also some allocated administration charges) we still find that on working account alone there was, by the end of 1962, a cumulative deficit of £208 million, the surpluses of 1948–55 being more than offset by the subsequent deficits even on operating expenses.

We can substantiate the contention that the trouble is not simply a financial mirage by looking at traffic trends. In a period during which Gross Domestic Product at constant prices has risen by over 40 per cent the amount of rail passenger travel has remained virtually constant and the rail transport of all the major freight categories has suffered not only a relative but also an absolute decline in volume (see Table 23).

(ii) Lack of investment

A second explanation is offered which also goes back to the transition from war-time to peace-time operation. The physical assets of

[1] See H. Wilson, 'Paying for the Railways', New Statesman, March 5, 1960.

TABLE 23

Rail Traffic Trends, 1948–60 (1948=100)

	1948	1950	1952	1954	1956	1958	1960	1962
Passenger miles	100	94·9	97·3	98·5	100·5	105·4	102·5	94·1
Net ton miles								
Merchandise	100	97·4	95·9	92·0	84·5	73·5	80·2	73·1
Minerals	100	101·2	104·8	101·8	105·0	85·9	97·4	72·4
Coal and coke	100	106·3	108·2	109·5	107·0	93·2	84·6	76·2

Note: The base figures for freight categories in 1948 have been adjusted to take account of a change in the basis of calculation after that date.
Source: B.T.C. Annual Reports and Accounts.

the railway system were seriously depleted during the war. Unfortunately the financial arrangement of a guaranteed net revenue did not permit sufficient reserves to be accumulated to deal with this backlog of replacement. Nor did the post war settlement make any specific provision in this respect, the assumption being that for a nationalized concern the necessary investment funds would be forthcoming. But in fact, the Labour Government, facing inflationary pressure in the post war period, deemed that investment in railways was an expendable item in the short run. It has been estimated that a total *disinvestment* in railways of some £440 million at 1948 prices (or about £650 million at 1960 prices) occurred between 1937 and 1953.[1] From 1954–60 gross investment in railways has amounted to £820 million at current prices but of this some £550–600 million was really only replacement necessitated by depreciation over the period, leaving net investment at current prices of only about £250 million. Hence it is argued that, at the end of 1960, railway investment in real terms still had some way to go before it even caught up with the accumulated disinvestment of the war and early post war years.[2] Admittedly, investment in roads was similarly restricted until the early 1950s; nevertheless the same did not hold for investment in road vehicles, which has been running at a very high level since the war, compared with investment in railway rolling stock. The conclusion that we may be tempted to draw from these facts is that more investment in railways would improve their financial situation and hence solve the problem.

This argument is clearly of some validity. Certainly in accepting the Modernization Plan in principle in 1955 the Government assumed that there must be many delayed but profitable investment

[1] P. Redfern, 'Net Investment in Fixed Assets in the United Kingdom', *Journal of the Royal Statistical Society*, Series A, Vol. 118, No. 2, 1955.

[2] D. L. Munby, 'Economic Problems of British Railways', *Bulletin of the Oxford Institute of Statistics*, Vol. 24, No. 1, February 1962.

projects in rail transport. The results, however, have been disappointing. Instead of being able to break even by 1961 or 1962 as envisaged in the original plan,[1] deficits on operating account of £87 and £104 million were sustained in these years. This can be attributed partly to adverse trends in some major traffics (coal, iron and steel) which could not reasonably have been foreseen; partly, however, it is the result of a failure both on the part of the B.T.C. and of the Government departments concerned to elucidate the fundamental aims of railway policy. For instance, as recently as May 1960, considerable confusion was shown in discussions of investment criteria before the Select Committee on Nationalized Industries. As D. L. Munby writes, ' . . . the Ministry and Treasury, in their discussions of the London Midland electrification scheme, do not seem to have realized that these were two different decisions involved: as to whether the line was so unprofitable that it should be closed, and as to what sort of investment should take place, if it was decided to keep it open'.[2] It was implied that the decision to keep a line open could be made not only in the interests of future profitability but also on social grounds. Though new investment might improve railway finances in either case, it would not be sufficient to ensure overall profitability unless the whole policy of railway management, including the decisions as to what services to provide, is directed towards profitability.

(iii) *Social service obligations*
We have already implied that confusion concerning its duties has inhibited the B.T.C. in seeking profits. Three roots can be traced to· the hesitant commercial attitude of the B.T.C.; firstly, those statutory restrictions involving supervisory machinery; secondly, ministerial interventions, both under the general directions clause of the 1947 Transport Act and by concealed pressure; and thirdly those self-imposed obligations which involved neither statutory restriction nor ministerial pressure.

(a) *Statutory requirements.* The 1947 Transport Act required the B.T.C. so to conduct its undertaking as to secure that revenue was not less than sufficient to meet those costs properly chargeable to revenue, taking one year with another. These costs were to include operating charges, allocations to reserve, servicing of capital, provision for depreciation and provision for redemption of capital. Unfortunately, the B.T.C. was not left a free agent to pursue these

[1] 'Proposals for the Railways', 1956, Cmd. 9,880.
[2] D. L. Munby, *op. cit.*, p. 27.

ends as best it could, but had to submit its charges structure to the Transport Tribunal, which had complete power over fares and charges until 1953 and power to fix maxima thereafter. The railways were well used to the procedure of submitting their pricing policy to the scrutiny of an independent authority. In the inter-war period the Railway Rates Tribunal had apparently fulfilled a similar function under the provisions of the 1921 Railways Act. In fact the Railway Rates Tribunal was generally prepared to agree to any suggested alteration in prices aimed at securing for the railways the 'standard net revenue'; as this target was never achieved during the inter-war years considerable freedom resulted.[1]

As the 1947 Act required the Transport Tribunal to abstain from any action which would prevent the B.T.C. from securing its proper revenue, and as profits were not forthcoming, similar freedom might have been expected in the post-war period. This has not been so for several reasons.

Firstly, there was no clear specification of what was meant by 'taking one year with another' so that the Tribunal could always refer implicitly or explicitly to the 'long term' when making decisions which would prove immediately deleterious to railway finances.

Secondly, there was the problem of delays and modification of both freight and passenger charges schemes. For instance, the freedom of charging within set maxima granted by the 1953 Act could not be implemented until the B.T.C. had devised a completely new charges scheme and had satisfied the Tribunal. A Draft Merchandise Charges Scheme was published by the B.T.C. in March 1955 but did not become fully operational until July 1957, a total delay of nearly four years. The scheme sought approval for a classification introducing the new concept of 'loadability' with maximum charge limits based upon traffic carried under adverse, though not extremely adverse, operating conditions. Despite the fact that the maximum charges applied for did not cover costs for some 10 per cent of existing freight traffic, involving cross subsidy to the extent of £53 million, and included no margin to cover any increases in costs which might be anticipated, the scheme was attacked on two main grounds. It was argued that the B.T.C. proposals were based on a 'maximum maximorum', aggregating adverse costs in each stage of its operations. The B.T.C. denied that this was so, though they did point out that in fact adverse conditions do tend to be associated in varying stages of operations. Alternatively, it was argued that the discrepancy between the maxima applied for and the normal cost was too great and that for a far higher proportion of traffic actual instead of maximum

[1] See Chapter 5.

charges should have been set, as envisaged in the 1953 Act for traffics where the railways retained an effective monopoly and flexibility was impractical or undesirable. Not only were the arguments prolonged, thus delaying the introduction of the scheme, but the outcome included average reductions in the maximum levels sought of some 14 per cent and the exclusion from the charges freedom of a much larger proportion of traffic than had been anticipated. Clearly the Tribunal's interpretation of the non-commercial obligations of the B.T.C. differed significantly from that of the B.T.C. itself.

Though the Minister was empowered to alleviate B.T.C. troubles by authorizing interim increases pending Tribunal decisions, this has not been effective in preventing the Tribunal from seriously affecting the B.T.C. financial position. The Select Committee on Nationalized Industries reported quite unambiguously on this point: 'In the past, both by the time taken to reach decisions and by the nature of the decisions themselves the Tribunal has severely cut down the Commission's earnings.'[1]

Thirdly, a more subtle point. Allocations to reserve are made by most large companies in order to finance new investment internally. As funds were in any case expected to be forthcoming for new investment from Government sources this allocation to reserve appeared to be unnecessary in the case of the B.T.C. Therefore charges were controlled at a level which did not permit the accumulation of reserves. But the B.T.C. need for a reserve was based not on the prospects of expansion but on mere self-preservation. Operating costs only included depreciation at historic cost and it was assumed that the difference between this and replacement cost depreciation would be provided in the accumulated reserve. Hence the Transport Tribunal control of charges was on a basis which implied constant erosion of railway assets.[2]

A less important statutory inhibition was imposed by the Transport Users Consultative Committees. The central committee and the eleven area committees included amongst their functions the consideration of closure proposals. Although the T.U.C.C.'s did not oppose closures where a strong case had been established to demonstrate that a facility was grossly unremunerative, they did expect the B.T.C. to make alternative arrangements for any passengers who might suffer undue hardship as a result of the closure. The T.U.C.C.'s have never been powerful but they have helped to slow down the trimming of unremunerative services and hence deleteriously affected

[1] Report of the Select Committee on Nationalized Industries—British Railways, June 1960, para. 410.
[2] Select Committee Report, 1960, *op. cit.*, Appendix 18, para. 3.

the B.T.C. financial position. These statutory restrictions were imposed to protect the public against a large monopoly operator. They were compatible with the overall financial requirement only on the assumption that profitable traffics were sufficiently extensive and assured to cross-subsidize those services which were particularly costly to provide. The changes involved in the 1953 Transport Act, whilst enabling the railways to offer competitive rates within the statutory maxima, still did not solve, or encourage the B.T.C. to solve, the problem of how to treat those unremunerative traffics for which there was no competition.

(b) *Ministerial interventions.* Apart from the effects of the machinery already described, the B.T.C. has also been restricted on occasions by direct ministerial intervention. In 1952 the Minister gave a formal direction to the B.T.C. instructing it not to increase passenger fares outside London after the passenger charges scheme had been accepted as reasonable by both the Transport Tribunal and (after reference to them by the Minister) by the Central Transport Consultative Committee. Again in 1956 the Minister refused to grant an interim increase in freight rates despite the recommendation of the Tribunal to allow such an increase; he also requested the B.T.C. to delay going to the Tribunal for an increase in passenger fares. The Select Committee on Nationalized Industries assessed the immediate cost of these interventions at between £15 million and £23½ million.

Similarly in wage negotiations the B.T.C. has from time to time been forced, by Government intervention, to incur cost which it might otherwise have evaded. In 1951 the B.T.C. was instructed to grant wage increases of 7½ per cent after a Court of Inquiry had recommended and the Commission had agreed to increases of 5 per cent. Again in 1953 the B.T.C. was forced by the Government to make concessions to the N.U.R. when under threat of strike action. Such intervention may be perfectly justifiable on the grounds that other restrictions make it undesirable for the B.T.C. to insist that wage increases are limited to what the Commission can afford to pay, but then it becomes illogical subsequently to castigate the Commission for the loss incurred due to the intervention.

It would not be just, therefore, to interpret the B.T.C. profit and loss account as rigidly analogous in meaning to the profit and loss account of a normal commercial enterprise during this period.

(c) *Self-imposed obligations.* Not all of the inhibitions were external to the B.T.C. however. In evidence to the Select Committee on Nationalized Industries Sir Brian Robertson pointed out that a purely commercial approach would have probably led to the abolition of all Sunday services and most commuter services, a course which he regarded as constituting total disregard of public need. Similarly it was established that even in new investment projects the Commission had, on occasion, chosen schemes on grounds other than the strictly commercial.

The Committee was moved to issue a clear reprimand over this. It reported ' . . . They (the B.T.C.) have been guided in a number of decisions by what seemed to them to be social needs as well as by what is economically wise. . . . If there are other considerations which make it desirable for members of the public to travel, or freight to be carried on some routes at prices below the cost, it should be for the Government and not the Commission to decide. . . . This confusion in judging between what is economically right and what is socially desirable has played an important part in leading to the situation in which the Commission now find themselves.'[1]

The implication of this reprimand, which the Government subsequently repeated in its White Paper of December 1960,[2] was that the B.T.C. had, through a misguided display of public spiritedness, failed to meet the clear and unambiguous requirements of the 1947 Transport Act.

It is quite arguable however that, if insufficient attention had been paid to the financial requirement of the 1947 Act, this was more the fault of the Government than of the Commission. Clause 4 of the 1947 Act states the general duty of the Commission 'to provide, secure, or promote the provision of an efficient, adequate, economical and properly integrated system of public transport . . .', and *only subsequently* that 'it shall so conduct that undertaking and levy such fares, tolls and charges as to secure that the revenue shall not be less than sufficient for making provision for the meeting of charges properly chargeable to revenue taking one year with another'. The wording of these requirements implies that achieving what would be regarded as an adequate service *was* compatible with financial viability. It does not specify, as critics have subsequently implied, that the financial aim should take precedence in the case of a conflict of aims. Thus, although the terms 'efficient, adequate, economical and properly integrated' would be very difficult to define, there is nothing illogical in interpreting this to be the overriding requirement, especially in view of the sequence in which the state-

[1] Select Committee Report, 1960, *op. cit.*, paras. 414–17.
[2] 'Proposals for the Reorganization of the Nationalized Transport Undertakings', December 1960, Cmnd. 1,248.

ments of duties lie. Even the amending clause in the 1953 Transport Act requiring 'due regard being had to efficiency, economy and safety of operation and to the needs of the public, agriculture, commerce and industry', remains vague in this respect. The relevant statutes did not envisage a situation where financial viability was incompatible with the standard of service traditionally regarded as adequate and hence did not provide terms of reference which would automatically resolve the problem. Indeed one might argue that 'adequacy' or 'due regard to public need' could not have been meant to be interpreted as 'provide such service as the public will render remunerative' or the clause would have been superfluous, this requirement having been explicitly included elsewhere.

Awareness of the magnitude of the problem developed slowly. The B.T.C. referred in their modernization plan[1] to two justifications for the provision of railway services, 'indispensability' and 'economy'. In describing an 'indispensable' service no reference was made to profitability. In the case of heavy bulk traffics it was considered against the broad national interest to raise charges to make them adequately profitable. Again in the reappraisal of the problem in 1959[2] reference was made to the conflict between the commercial and the public service function of railway operation. On neither occasion did the Minister to whom the reports were submitted take up this fundamental point despite the responsibility for overall supervision of major questions of policy laid on him by Clause 4 of the 1947 Act. Not until the Reorganization White Paper of December 1960 did the Government explicitly recognize, and take steps to deal with, the problem. Throughout the 'fifties not only were the Commission inclined to pursue ends other than the strictly commercial in their interpretation of statutory requirements but the Government, both by acts of commission and omission, gave them every encouragement to do so. Hence any brickbats should be directed at least equally at the Government and the B.T.C.

(iv) Unfair competition

The cry of 'unfair competition' has been raised very frequently since the end of the First World War. There are three main strands of complaint. Firstly, and most commonly, is the 'track cost' argument that whereas charges for rail track are related to cost, charges for use of the roads are not, particularly in that no servicing is required on the capital involved. We have dealt with this argument in detail

---

[1] 'B.T.C. Proposals for the Railways', October 1956, Cmd. 9,880.
[2] 'The British Transport Commission—Reappraisal of the Plan for the Modernization and Re-equipment of British Railways', July 1959, Cmnd. 813.

elsewhere.[1] Suffice it to point out here that since 1953 the railways have been free to offer competitive charges if they so desired. That is, they have been able to reduce their quotation as far as short run marginal cost, excluding amongst other things any element of track capital cost. Even on this basis their traffic has declined. For the track cost argument to be maintained in its crude form in face of the evidence it is necessary to assert that the 'true' track cost per unit carried is higher for road transport than for rail and that thus competition on the basis of marginal private cost is favourable to the road agencies. Even in principle it is difficult to envisage what is meant by 'true' cost in this context for apart from the problem of valuing the capital involved there is a problem as to the spread of traffic over which the capital costs should be allocated. Neither by intuition nor by calculation with the only available figures, however, is it clear that road transport is more capital intensive in respect to track capital per output unit than rail.[2] To criticize the 'track cost' argument in its crude form is not, of course, to argue that comparability currently exists; indeed, the arguments subsequently adduced for an efficient use of road resources[3] are based on a re-interpretation of what are the 'real' costs of road use at the margin.

A second type of unfair competition argument says that, although the railways have freedom to charge down to marginal cost in the short run, in fact it is the road operator who is more likely to pursue such a policy. Consequently, even though there are bankruptcies in road haulage and the individual operators concerned do not benefit, nevertheless the overall result will be to transfer traffic from rail to road, cause some rail services to appear unremunerative and hence to be withdrawn even though in the long run in real terms the rail facility is more economic. This argument might be valid if the

---

[1] See Chapter 9.

[2] Track capital/output ratio for freight by road and rail, 1952.

	Road	Rail
Output ('000 m. ton-miles)	18·8	22·4 (a)
Track capital (£m.)	556	1,008 (b)
Track capital/output ratio (£ per '000 ton-miles)	29·6	45

Sources: (a) Transport of Goods by Road, H.M.S.O., 1959.

(b) These figures are from Redfern, op. cit. They consist of depreciated replacement value estimates for road and for all rail asset less rolling stock. The estimated value will depend on the age structure of the assets. The whole calculation is tortuous but at least shows that there is no obvious advantage accruing to road transport in this respect.

[3] In Chapter 12.

M

discrepancy between short run and long run marginal cost was greater for the road than the rail operator. This is not so. Consequently road operators are no more likely to charge uneconomically low rates than are rail; even if they do, the railways are free, to defeat them by also charging down to short run marginal cost. Hence any disadvantage in this field is more likely to indicate commercial ineptitude on the railways' part than a structure artificially biased against them.

A third ground of complaint is that the railways suffer a creaming of their traffic, losing the most remunerative and retaining only that which is unprofitable. Loss of profitable traffic due to price competition has been dealt with in the last paragraph. Retention of unprofitable traffic has been unavoidable wherever the maximum charge was below the cost of carriage by other agencies. Such restriction on the railways' ability either to charge a suitable rate or to shed such traffic amounts to a social service obligation imposed on the railways.

Hence the 'unfair competition' argument is valid only in so far as it can be demonstrated that it has not been possible for road-rail competition to proceed on the basis of marginal real costs and that the bias has been anti-railway.

THE ORIGINS OF THE BEECHING PLAN

Our investigation has revealed that an over-extended capital structure, delayed replacement and new investment, social service obligations and some incomparability between the commercial freedom of competing agencies have all contributed to the financial plight of the railways. We have generously argued that these inhibitions have largely been beyond the power of the B.T.C. to alter and that the losses should therefore to a certain extent be attributed to the institutions rather than to the railway management. Clearly, however, this is not the whole story. The railways themselves are to blame in so far as they have been performing services wastefully or performing the wrong services. The Beeching report[1] is a clear exposition of the commercial potentiality of the railways. The plan represents the culmination of a process of trial and error whereby it has become clear that even with modern equipment many of the transport services which had traditionally been performed by the railways could no longer economically be allocated to them.

[1] British Railways Board, *The Reshaping of British Railways*, H.M.S.O., March 1963.

The antecedents of the new policy can be usefully traced back as far as 1953. The reintroduction of competition by the Transport Act of that year was accompanied by the determination to make up the arrears of capital investment in railways. Consequently the plan for the modernization and re-equipment of British railways evolved during 1954 was launched early in 1955 with the full blessing of the Government which described it as 'courageous and imaginative'. It included a number of fundamental technical changes such as the replacement of steam locomotion by diesel or electric traction, replacement of outdated passenger stock, new traffic control and telecommunication systems and a drastic remodelling of freight services, including the introduction of larger mineral wagons, continuous braking, and the resiting and reconstruction of many goods depots. This technical modernization was not the only aspect of the plan, however. In the 1956 review[1] the B.T.C. stated four ways in which they were attempting to achieve a profitable basis of operation, namely:

(i) Modernization,
(ii) Elimination of unprofitable services.
(iii) Increased productivity.
(iv) The further relaxation of statutory obligations, particularly in respect to charging.

Unfortunately the plan did not fructify as rapidly as had been anticipated. A sudden decline of activity in the heavy industries in 1958 seriously affected receipts and, apparently, expectations for the future. The 1959 reappraisal[2] argued, however, that the plan had been soundly conceived and emphasized the need for acceleration so that the commercial advantages could be reaped earlier. Again, in accepting the B.T.C. reappraisal, Government silence may be taken to indicate assent, or at least absence of fundamental dissent.[3]

The turning point in Government attitude to the railways came in 1960. In July the Select Committee on Nationalized Industries published its report on the railways and during the autumn the Special Advisory Group set up by the Minister to advise him on the nationalized transport structure also reported. We can only infer the contents of the latter report from the contents of the reorganization

[1] Cmd. 9,880, op. cit.
[2] Cmnd. 813, op. cit.
[3] Not until January 1961 did the Government publicly declare that they did not endorse the reappraisal. See Hansard Commons Debate, 1960–61, Vol. 633, Cols. 624–6, 630.

White Paper[1] which claimed to be largely based on it, for the report itself was never published.[2]

The Select Committee, however, made it clear that the problem was not simply a question of minor alterations but of major changes of policy, finding fault with the commercial methods of the B.T.C., its organization and its aims.

Three major criticisms of B.T.C. commercial methods were made; of the slowness in developing a satisfactory costing service, of the delay in incorporating a regional accounting system, and of failure to account and estimate with sufficient precision the effects of the modernization plan. On the first two points the B.T.C. replied indignantly.[3] Railway costings, they asserted, were particularly difficult to achieve and the B.T.C. were well to the fore in this field. As for regional accounting, they had already instructed a firm of accountants to advise them on how regional accounts could best be achieved. In fact, as the accountants subsequently reported, separate regional accounts cannot have the same meaning as the Consolidated Accounts for British Railways. With 45 per cent of all traffic terminating in a region other than that of origin, separate financial accounts could only be prepared on the basis of arbitrary divisions of receipts. Indeed such accounts might encourage regional management to concentrate on seeking traffic which would benefit the regional account rather than the Consolidated Account.[4]

The main criticism of the organization of the Commission concerned the absence of a separate responsibility for the railways which clearly posed problems quite different from those involved in the other B.T.C. activities. It is interesting to note that despite the recommendation for regional accounting the Select Committee did not recommend any other extension of regional autonomy, considering that the existing extent of central co-ordination and regional decentralization of management was just about right.[5]

[1] Cmnd. 1,248, *op. cit.*

[2] In answer to a question in the House on December 21, 1960, the Minister of Transport, Mr Marples, refused to alter the decision not to publish the report. He argued that when the Stedeford Group was set up it was felt that more evidence could be obtained from interested parties, including the T.U.C., if it was to be confidential. To publish it after giving such assurances of secrecy would be a breach of confidence. See *Hansard Commons Debate*, 1960–61, Vol. 632, Col. 1,301.

[3] Special Report from the Select Committee on Nationalized Industries—British Railways (Observations of the B.T.C. and of the Minister of Transport), H.C., 163, 1960–61.

[4] No further reference has been made to complete regional self accounting either in the 1962 Transport Act or the Beeching plan. One must presume, therefore, that the suggestion has been quietly interred.

[5] Select Committee Report, 1960, *op. cit.*, paras. 20–60.

Finally, in reviewing the aims of the B.T.C. the Select Committee recommended that to solve the 'confusion in judging between what is economically right and what is socially desirable' the B.T.C. should act solely in pursuit of profits and the Government should act openly where it considered that modifications of the resulting B.T.C. policy needed to be made on grounds of national economy or social needs. Moreover, the B.T.C. should be compensated for any such interventions so that their commercial and economic efficiency might more fairly be judged, as is the case with private enterprises, by reference to the profit and loss account.

On the basis of the reports of the Select Committee and the Special Advisory Group a plan for the reorganization of the nationalized transport undertakings was published in December 1960[1] and subsequently implemented by the 1962 Transport Act. Radical changes were to be made in three respects, namely:

(i) Reorganization of the structure.
(ii) Reconstruction of the finances.
(iii) Extension of commercial freedom.

The result is a Railways Board, specifically devoted to railway matters, carrying a largely suspended liability to interest payments looking suspiciously like a Treasury held equity interest, free to treat the problem on a purely commercial line. For the rest of this chapter we shall be considering, in the light of the Beeching report, the prospects of such a railway undertaking.

THE RESHAPING OF BRITISH RAILWAYS

Rail transport depends upon the exclusive use of a specialized route system. Maintenance of this special track and signalling facility by itself accounted for 20 per cent of total costs in 1961 (£110 million out of £561 million). Not only the track, however, but also much of the terminal and handling facility is fixed in the short run and indivisible once the decision has been made to provide a service of a certain maximum capacity. The use of this high cost track with its very rigorous traffic control system gives rail transport three distinct advantages over other agencies:

(a) High capacity units of carriage can be used,
(b) Dense flows can be accommodated,
(c) High speeds can be achieved with perfect safety and predictability.

[1] Cmnd. 1,248, op. cit.

Hence we might expect the art of operating a railway system to consist of providing such facilities and securing such traffics as would enable a dense flow of high capacity units to be handled at high speed, in order to spread the overhead to the fullest possible extent.

Unprofitable railway operation in the past has been largely due to the acquisition or retention of traffics which do not exhibit these physical characteristics. Even traffics moving sparsely in small loads at low speeds may, however, be able to earn enough revenue to justify the heavier incidence of fixed cost per unit. The crucial requirement is that the sum total of the traffic on a route should cover not only the costs which are fairly directly variable with the amount carried but also the system cost of that route. However large or small the unit of service under consideration, if the net revenue attributable to it (i.e. the revenue that would be lost if it were to be withdrawn), is less than the costs immediately escapable by withdrawing the service, then profitability will be best served by withdrawal. This holds equally for an extra coach on a train, an extra train on a line, or for the line as a whole. Of course, some subtlety may be required in the assessment of net revenue. A branch line may feed traffic on to, or receive traffic from, a trunk line. If the traffic would not use the trunk service were the branch feed line to be closed then the net revenue which should be attributed to the branch line would be all receipts from traffic originating or terminating in the branch lines less the directly escapable cost on the trunk line of the traffic concerned. Or, to take another example, many passengers may use a given half hourly service because they are assured of its regularity. At certain times of the day the number of passengers using the service might not appear to justify the continuation of particular trains. However, if some trains were withdrawn the service would no longer have the advantage of completely predictable periodicity and some passengers using the service at various other times throughout the day might be lost. Again net revenue might be considerably higher than the immediately apparent net revenue attributable to the train.

The revenue obtainable depends of course on the quantity of traffic forthcoming at the price offered. Hence the decision concerning what facilities to operate really requires that alternative price levels are considered so that services are only withdrawn when it is quite impossible to make them pay by variation of charges. Similarly, some of the more radical suggestions for the reduction of operating cost ought to be tried out before any nice decisions are made. For instance, the wastages involved in the traditional forms of station staffing might be eliminated[1] and the traditional engineers standards for track maintenance might be critically reviewed in the light of current operating conditions.[2]

There are, however, many services for which the disparity between net revenue and escapable costs is so large that no conceivable commercial or operating practice could make them pay. These unprofitable propositions are part of a general pattern; certain categories of traffic, widely defined according to the physical nature of the traffic and the conditions of carriage, are generally suited to rail transport whilst others are not. We shall now go on to describe this pattern.

Passenger services

Passenger services were responsible for one third of railway gross receipts in 1961. Though a total loss of over £100 million was estimated for passenger services as a whole, not all types of service were equally unprofitable.

(i) *Fast and semi-fast inter-city routes*. Though, after the allocation of indirect costs, a deficit was shown on this type of service, at least a substantial surplus was shown over the direct costs of operation. In this sphere rail transport can be highly competitive with road transport both on grounds of comfort and on grounds of speed. Departure or arrival times carefully chosen in relation to the needs of the communities involved frequently permit the achievement of high load factors so that the cost per passenger mile is kept relatively low. For instance, particular attention to the demands of business travellers for speed and comfort may enable most of the radial services from London to the provincial centres to be operated profitably.

Not all trunk routes have such good prospects. Competitive railway building in the nineteenth century produced much economically unjustified duplication of facilities. Many large towns have more than one central railway station and are served by parallel routes to the metropolis. Elimination of such excess capacity will produce cost savings without any correspondingly serious revenue losses.

Even on profitable trunk routes seasonal traffic may produce

---

[1] See R. Shervington, 'Reducing Station Costs', *British Transport Review*, Vol. 6, No. 6, August 1962. Also the *Report of the Central Transport Consultative Committee for 1962*, p. 12.

[2] See D. L. Munby, 'The Reshaping of British Railways', *Journal of Industrial Economics*, Vol. 11, No. 3, July 1963.

both commercial and operating problems. Though it appears to be profitable by virtue of the high train loadings obtained, in fact it gives rise to extremely heavy capital costs, much of the stock used for peak services being idle for the rest of the year. For instance, only 5,500 of the 18,500 gangwayed coaches allocated to fast and semi-fast services in 1959 were used all the year round. 8,900, or 48 per cent of the total, were used only for high peak services; 6,000, or nearly a third of the total, were used on not more than 18 occasions during the year. Fortunately, peak traffic is being eroded by road competition more rapidly than other traffic. Whereas in 1951 July traffic exceeded the annual average monthly traffic by 96 per cent, by 1961 this difference had fallen to only 47 per cent. The seasonal peak may in future be further controlled by fares policy, including some differential charging of high peak travel through a system of seat reservations.

(ii) *Suburban services*. In 1961 suburban services did not even cover their direct costs. The total loss incurred was estimated as about £25 million, though in London, which produced 86 per cent of sub-urban revenue, direct costs were covered and only a small deficit in-curred on total cost. The general problem arises because of the intensity of the peak load, which, measured in London over half an hour, is twelve times the average over the day as a whole. Conse-quently stock can only be used at 8–10 per cent of capacity for most of the day. Despite the heavy capacity costs involved by this peak demand most of it is satisfied at fares less than the standard rate, through the means of season ticket rates. The physical difficulties of commuting to London other than by rail makes demand relatively inelastic so that the abolition of the cheap rates, or even an overall increase in fares, would increase the revenue of the suburban services. This has been prevented by the Transport Tribunal control of fares in the London area.

Outside London the losses on suburban services are beyond redemption by fares policy alone. In smaller conurbations road transport is a more viable alternative. Consequently demand for rail transport is more elastic. Fare increases would not increase revenue significantly, if at all; fare reductions might attract new traffic sufficiently to increase revenue but only at the cost of in-creased peak capacity. In the absence of any external offer to subsidize these services on the grounds that the alternative of new invest-ment in roads is even more expensive, commercial considerations would suggest that they be withdrawn.

(iii) *Stopping train services.* Services linking smaller communities are the least viable of all. It has been argued that cutting out the least patronized trains might, by reducing costs more than revenue, improve the position. Unfortunately the lower the level of utilization of track the greater the proportionate significance of overhead costs per unit. Thus, though some savings may be achieved by reductions in service, the increased incidence of overheads would necessitate greatly improved loadings on the remaining services if the overall situation were to be improved.[1] Nor would the manipulation of fares promise much relief. In this case demand is likely to be relatively inelastic so that price reductions would reduce gross revenue; it is unlikely however to be sufficiently inelastic to permit the converse policy of price increases to revive the financial fortunes of grossly unremunerative lines. The prospects for rural services are therefore dim.

This bleak outlook may be improved slightly in two ways. Firstly, if a line would be kept open in any case for freight traffic then the escapable cost of a passenger service is confined to movement cost plus any extra system cost due to passenger traffic. This will permit a lower passenger density to be economically feasible than would otherwise be so. Secondly, if traffic fed to or from main lines will be lost to rail as a result of the closure of a branch line the whole of the net revenue involved should be imputed to the branch line in a closure consideration. Interdependence of services seems to have been taken into account in the Beeching proposals only in respect of outgoing traffic; no information appears to have been collected in this respect for traffic received. The extent of traffic losses consequent upon a closure will depend to a certain extent on imponderables such as how well the alternative forms of transport used are co-ordinated with the trunk rail services. Nevertheless, the omission may be serious, especially in the case of predominantly receiving areas, such as holiday resorts.

Freight traffic

The rail network was mostly constructed before the development of either an adequate road system or efficient road vehicles. Railways were therefore extended as far as possible to minimize the need for ancillary haulage by road. Many tiny stations and depots were constructed to handle traffic originating in small consignments. This multiplicity of railheads, together with a statutory control of charges preventing discrimination in favour of large consignments, perpetuated the wagon-load as the unit of movement. This required a

[1] This point is taken up again in greater detail in Chapter 13.

great deal of marshalling causing extremely slow wagon movement. Moreover, relatively infrequent collection and delivery of wagons at the smaller depots and stations produced long terminal delays. Terminal delay and slow wagon movement together necessitated an enormous wagon fleet to provide the services required. Such methods were extremely costly both in the utilization of labour (due to excessive handling) and of capital (due to the inefficient utilization of wagon fleet). Moreover, because each wagon must be capable of coupling to any other wagon (the principle of compatibility), improved design in rolling stock was greatly impeded. Traditional methods of handling freight traffic have thus been almost diametrically opposed to the principle advantages of rail transport which lie in catering for large consignments at high speed for long stages on heavily used trunk lines.

The story can be spelled out in greater detail by considering the three major classes of traffic, of which, as Table 24 shows, coal is the most important in respect of tonnage carried, gross receipts and net receipts.

TABLE 24

Profitability of Freight Traffic, 1961

Class of Traffic	Tons	Receipts	Margin of receipts over	
			Direct cost	Full cost
Coal	145·7	108·3	24·8	2·8
Mineral	54·3	44·5	7·6	− 3·7
General Merchandise				
Wagon Load	34·4	64·8	−31·8	−53·8
Sundries	3·8	38·0	−13·5	−21·3
	38·2	102·8	−45·3	−75·1

Source: The Reshaping of British Railways, p. 26.

(i) *Coal.* In 1961 some 89 million tons of coal or 61 per cent of the total traffic in the class was still transported in wagon-load consignments. Of this, 54 million tons went to sidings and 35 million tons to stations. 3,869 separate stations dealt with the latter, each station on average distributing by road over a 2½-mile radius. By concentrating on larger depots yet only increasing the road haul to a 10-mile radius the number of distribution points could be reduced to 250. This would enable the depots to be fed by train-load consignments and would probably justify efficient mechanization of the depots. Costs of rail transport would be almost halved by these favourable handling conditions so that despite the increased cost of the final stage road distribution a net benefit could accrue to both consumer and railway. With this in mind it is proposed to reduce the number of small stations open for coal traffic.[1]

Most of the 39 per cent of coal traffic which currently travels in train-load consignments goes either to large industrial consumers or to the Central Electricity Generating Board. Due to concentration of both heavy industry and electricity generation on the coal fields the average length of haul is low. Estimates of coal consumption suggest that the proportion consumed by C.E.G.B. will increase rapidly in the next five years from 28 per cent to 38 per cent of total coal consumption at the expense of other uses of coal. Consequently ton mileage by rail and gross receipts for coal carriage could well decrease. Nevertheless net revenue might increase due to the more favourable operating conditions for the large consignments involved.

Expense might be avoided further by a more efficient use of wagon stock at the colliery. Currently wagons spend an average of two days in colliery sidings acting as free, or almost free, bunkers. It is estimated that the cost to the railway of wagon use lost in this way is about £11 million per annum, whilst the N.C.B. pay a total demurrage charge of only £1 million per annum.

Coal transport, despite being currently profitable in total, is then still susceptible to considerable improvements which should benefit both railway accounts and the consumer.

(ii) *Mineral traffic.* Of the 54·4 million tons of mineral class traffic carried in 1961, 43 per cent moved in full train loads. For some commodities, notably iron ore, efficient terminal equipment, special wagons and large trains are already the rule with the result that freight rates are low but still yield profits to the railways.

(iii) *General merchandise.* Of the 34·4 million tons in this category only 6·5 million or 19 per cent was carried in full train loads and the category as a whole is unprofitable. A further 3·8 million tons of sundries were consigned in less than wagon-load quantities. This gives rise to very poor wagon loadings and often requires transhipment which makes carriage extremely expensive.

The variation of cost with size of consignment has already been noted; a similar variation operates according to terminal handling conditions. Any traffic either road collected or road delivered by railway operated vehicles is carried at a loss even on direct costs. This is partly because of the physical inefficiency of terminal handling methods and partly because of the very low utilization which is made

[1] *The Reshaping of British Railways, op. cit.,* p. 33.

of the excessive number of terminal delivery and collection facilities. Similarly even station traffic involving transhipment into customers' vehicles is likely to be unprofitable on the same grounds of poor utilization of station facility. By far the most favourable terminal conditions are to be found in sidings traffic because no transfers from vehicle to vehicle are involved and this type of terminal facility is less costly to provide. There is, moreover, a close association between optimal terminal conditions and loading characteristics.[1] Traffic involving the lower cost terminal facilities frequently possess higher wagon and train loadings. Hence the declared railway policy is to reduce the number of stations handling freight, and to concentrate on methods of carriage which avoid expensive terminal or transitional handling by the railways themselves. This can be done in two ways:

(i) By encouraging through-train working.
(ii) By devising a method of handling less than train-load consignments with a minimum of transhipment. This involves the concept of the liner train which is relatively new to British railways.

The liner train
The proposed liner trains would provide a combined road-rail service for long distance freight haulage. Containers loaded at the point of origin would be hauled to the liner train depots by road feeder vehicles. They would be transferred by crane on to service trains of permanently coupled flat wagons. Similarly, unloading and delivery would be in container bulk. In this way a fast scheduled service could be provided over a network of major routes without expensive reloading from road to rail container and vice versa, eliminating expensive marshalling movements of railway wagons (and the physical wear on the consignments caused by shunting shocks), simplifying documentation and almost totally eliminating pilferage risks. Estimates[2] suggest a door to door cost advantage over road for hauls longer than 100 miles together with higher speed and at least equivalent quality of service to that offered by road haulage.

For a proposed network of services involving fifty-five terminals and depots a current traffic potential of about 30 million tons per annum is estimated. Half of this consists of traffic currently carried by road which would be won for road-rail combined service, 12 million tons consists of unremunerative wagon-load traffic at present

[1] Though the average length of haul for traffic in this category is relatively low.
[2] *The Reshaping of British Railways*, op. cit., Appendix 4.

flowing through stations and 3 million of sundries traffic currently flowing between main centres.

Both the concentration on through-train movement and the liner train proposals represent an intention to concentrate railway service on those aspects in which it has physical advantages over other agencies, namely safe handling of bulk flows at speed over long distances. We might therefore expect such a reconstituted rail system to regain some of the traffic which has been lost due to the relative inefficient concentration on the wrong traffics and to restrictions on commercial policy that have been operative hitherto. The B.T.C. traffic survey revealed 13 million tons of potential sidings traffic and 16 million tons of potential liner train traffic which, by vigorous selling and soundly based operating policy, could well be regained.[1]

This additional traffic barely covers the quantity which will be lost by withdrawal of facilities. As the proposals aim to secure a more intensive and productive use of both labour and equipment, redundancy must therefore be expected in both categories. In equipment the reduction of the post war period is to be accelerated. For instance, whereas in the fourteen years from 1948 to the beginning of 1962 railway wagon stock had decreased by 22 per cent, in the following two years alone a decline of a further 20 per cent was anticipated. The implications for employment of labour were not so clear, but the plan appears to have envisaged a decline of about 70,000 or 15 per cent, over a period of about five years compared with a decline of 26·9 per cent over the previous fourteen years. This did not, on the face of it, appear to be a very substantial acceleration. It was hoped that with the high annual turnover of labour which the railways experience (80,000 per annum) that the reduction could be secured primarily by control of recruitment. Problems would arise of course. Natural wastage is not always suitably geared to redundancies, either geographically or by occupation, and there may be some employees for whom no suitable alternative railway occupation is available. In view of this, the established redundancy agreements were revised. Persons downgraded were to retain their former rate of pay for up to five years, substantial removal grants were offered for those forced to move home and for those discharged there were to be resettlement payments and supplemental unemployment pay. Thus it was hoped to minimize the problems of transition which would arise as a consequence of the economy of labour resources.

The future of the railways, as envisaged by the Beeching plan, thus lies in more intensive use of both capital and labour. This

[1] *Ibid.*, p. 41.

implies a conscious abandoning of services for which rail transport is not naturally suited, and, at the same time, a serious attempt to translate the potential technical advantages of rail transport for other traffics into real commercial advantages. In this way a more economical division of function should be attained, not only through a correct allocation of traffic between agencies but also, where appropriate, by combining the advantages of different forms of transport.

CHAPTER ELEVEN

Other Transport Agencies

Our analysis of British inland transport would be incomplete without some assessment of the present and likely future significance of transport other than by road or rail. We shall therefore briefly consider the character and prospects of transport by air, coastal shipping and pipeline.[1]

AIR TRANSPORT

Air transport in Britain is confined primarily to the field of public passenger transport. Freight shipment by air is only economic for valuable consignments required with great urgency and for the least accessible parts of the country. Similarly private air passenger transport is beyond the means of most individuals though a number of large firms now operate executive aircraft. The Business Aircraft Users Association, formed to facilitate such flying, had seventy-six members by mid 1963 and there were about 200 business aircraft on the register. In such a small country the scope for either freight or private passenger transport by air is limited.

In contrast, public passenger travel on scheduled air services is not only currently of some significance but is also expanding rapidly (see Table 25). Though air transport is still responsible for less than 1 per cent of all domestic passenger travel it may have considerable impact on individual trunk routes.

TABLE 25

Domestic Air Services in the U.K. Year ending October 1962

	Total	B.E.A.	Independent	B.E.A. %
Seat miles offered (million)	1,050	803	247	76·5
Increase on previous year (%)		+20·8	+26·7	—
Passenger miles flown (million)	694	539	155	77·6
Increase on previous year (%)		+17·5	+20·2	—
Passenger load factor (%)	66·1	67·1	62·9	—

Source: Ministry of Aviation departmental statistics.

[1] Inland waterways are omitted on the grounds of their negligible significance. In 1962 they accounted for less than $\frac{1}{2}$ per cent of freight ton-mileage. For a description of the network, its traffic and its prospects see Report of the Committee of Inquiry into Inland Waterways, 1958, Cmnd. 486.

Within the category of domestic air services, however, many routes compete with sea transport (e.g. services to Northern Ireland, Channel Islands and the Hebrides). Though such routes do also eliminate a certain amount of inland surface mileage, the potential for the substitution of air for inland surface transport is much greater in the case of routes in direct competition with through road or rail facilities.

These purely internal services account for over half of B.E.A. domestic traffic and about a quarter of the independents' traffic. Moreover, in the case of B.E.A., this is also the most rapidly expanding business with an increase of 25–30 per cent per annum over the last three years. Though prior to 1960 the independents were only able to operate internal scheduled services as associates of B.E.A., on routes which B.E.A. had not decided to cover, since the Civil Aviation Act of that year they are, in principle, able to compete on routes worked by B.E.A. Hence recent trends suggest an increase both in the independents' share of domestic air transport and in the share of air transport as a whole in domestic passenger traffic.

What factors will determine the nature and the extent of this expansion of air transport?

(i) *Public control.* Firstly we must recognize that air transport is one of the most highly cartelized industries in the world. The individual operator is either limited by agreements which he has voluntarily entered, or, if he is a small operator outside the cartel, he is controlled by his national Government so that he does not upset the stability engendered by the cartel. It is impossible to understand the structure of the industry except within the context of these restrictions.

The main reason for such control is the close connection which exists between civil and military air operation involving questions of military strategy and national prestige.[1] For instance, when B.E.A. wished to purchase a new type of aircraft the project would first be discussed by an interdepartmental committee, the Transport Aircraft Requirements Committee, in an attempt to reconcile the requirements of civil and military aircraft users.[2] Indeed, the non-statutory links between the Minister of Aviation and the nationalized air corporations have been so close that the Select Committee on Nationalized Industries suggested in 1959 that they were resulting in an undesirable diminution of the authority and sense of responsibility of the Boards, to the detriment of commercial efficiency.

[1] See L. Foldes, 'Domestic Air Transport Policy', *Economica*, Vol. 28, 1961.
[2] See the Report of the Select Committee on Nationalized Industries, 1959—The Air Corporations, pp. 10 ff.

Not only the nationalized air corporations are subject to control however. No public air service of any kind may be provided without a licence. Until the passing of the Civil Aviation Act, 1960, these were obtained from the Minister, acting on the advice of the Air Transport Advisory Council; since then the decisions are made by the ostensibly independent Air Transport Licensing Board, though appeal is to the Minister who thus retains power of review. In fact, overall policy is no less effectively dictated by the Minister under the new structure than under the old.

(ii) *Speed*. The demand for air transport depends heavily on the extent to which it can achieve significant time savings compared with surface agencies. In actual speed of movement the aeroplane has a clear advantage over the motor-car or railway train. But this advantage is frequently counterbalanced by the fact that surface transport routes are from city centre to city centre whereas airports are normally at some distance from the city, thus necessitating ancillary transport by road or rail and producing terminal time wastages. The shorter the air stage the greater is the relative importance of the terminal delay. For this reason the advantages of air travel in terms of time savings do not become significant until the stage length approaches 200 miles where two large conurbations are being connected. The development of an economical, high capacity, vertical take-off craft might upset this conclusion by enabling air transport to take place from city centre to city centre but as yet there is no sign of such a craft becoming commercially feasible.

(iii) *Comfort and convenience*. The standard of comfort varies with the type of aircraft used. Restricted entry and the absence of price competition in international service during a period of rapid technological advance has led to intense competition in quality of service. Aircraft have become outdated for international flying long before they were physically worn out with the result that there has been a good deal of high standard capacity available for domestic flying. This might well be viewed as a form of cross-subsidy of domestic air transport if aircraft were amortized over a period considerably less than their physical life and subsequently either sold at a relatively deflated price or used on internal services on the basis of this artificially low capital cost. Though craft built for long distance work are not completely interchangeable with those designed for shorter stages and hence there is a limit to the usefulness of many craft for domestic services, nevertheless it is likely that for some time to come

N

it will be possible to get good value for money in second-hand aircraft. In particular, independent companies may be able to provide a higher standard of service at a given price than would be possible if they had to order new aircraft of their own.

Convenience is predominantly a function of regularity of service. As units of relatively low capacity may be operated it appears likely that more convenient scheduling will be possible for air than for rail transport for a given total flow of traffic. This argument should not be over-emphasized however, for the cost per seat mile of air transport is lowest for those aircraft with relatively high capacity. For instance, B.E.A. estimated that in 1961–2 the average cost per seat mile on a Pionair Leopard (capacity thirty-two) was, at 8·4d, just over twice that of a plane twice its size, the Viscount 800 (capacity seventy) having an average cost per seat mile of 4d.[1] On sectors where the total demand is not great regularity of service can therefore only be provided at considerable cost either through the use of a small aircraft with a high cost per seat mile or through poor utilization of a craft of higher capacity. Thus the optimum convenience and cost combination is likely to be achieved on stages with a high total traffic potential such as the London-Glasgow stage.

(iv) *Costs*. We have shown that the cost per seat mile is dependent on the total demand on a particular route. The converse is even more obviously true; the demand for air transport, as for any other commodity or service, is dependent on its cost. We shall now consider the costs of air transport in greater detail.

In many respects the cost structure for air is very similar to that for rail transport. For instance, very little cost is attributable to the individual passenger in the sense of being escapable if that passenger did not travel. With capacity normally well in excess of demand the costs escapable if one passenger were to withdraw consists of little more than the coffee or glucose sweets that he would have consumed in the course of the flight.

The first major watershed in costing comes at the minimum unit of service provided, in this case the flight. B.E.A. accounts assemble under the head 'variable costs' all those costs attributable to particular flights in the sense that they could be escaped if a single flight were withdrawn. This category includes fuel, oil and operating materials, meals and accommodation of crew and cabin staff, handling fees, landing fees, commission payable, the costs of chartering aircraft where this has been necessary, and certain engineering costs which relate to the routine maintenance and inspection which has to be

[1] B.E.A. Annual Report and Accounts for 1961–2, Appendix 10.

carried out before any flight. For B.E.A. domestic services variable costs accounted for 40 per cent of the total.

The next category of costs shown in the B.E.A. accounts are called 'allocated and apportioned costs' and are those considered to be directly traceable to particular aircraft, though not, except in an arbitrary fashion, to particular flights. This is a rather heterogeneous category. Some of the costs, such as aircraft standing charges and the pay of the crew and cabin staff attached to the unit of capacity, can be directly traced to the individual unit and hence allocated to that unit; others, such as engineering, base labour and overheads, station costs and advertising, are not so directly traceable and have to be apportioned between aircraft according to some arbitrary criterion. In terms of escapability it is clear that the allocated costs would be escapable if the unit of capacity were eliminated whereas with the apportioned costs this does not seem to be the case. However it does make sense to separate the apportioned costs from the remaining category of pure overhead, consisting of centralized services administration and sales overheads because they can be directly attributed to particular routes in the sense that they would be escapable if the route were to be abandoned. Allocated and apportioned costs account for about 50 per cent of the total and the pure overheads for the remaining 10 per cent.

In costing a particular service B.E.A. use the plane as the basis and apply costs to it. Allocated costs are distributed in proportion to the scheduled use of the aircraft over a year, apportioned costs being distributed between aircraft in proportion to the gross weight of craft in the fleet. The first of these accounting conventions stimulates high aircraft utilization; the second encourages both high utilization and the choice of craft with a high payload to gross weight ratio.

The average cost per seat mile on any particular service will depend on a host of factors including stage length, type of aircraft used, intensity of aircraft utilization, frequency of service, load factor and mail pay. As stage length increases the proportionate significance of landing charges decreases. Moreover, both the block speed and the proportion of crew time spent in flight increase, thus reducing cost per mile flown. The cost per passenger mile, as opposed to the cost per aircraft mile, will continue to fall as stage length increases up to the point where more fuel and less passengers have to be carried to achieve the stage length. This optimum stage length, like the average cost per seat mile, will vary between types of aircraft. Costs per seat mile also fall fairly steadily as the annual utilization of an aircraft rises up to about 3,500 flying hours, after which the maintenance costs rise so rapidly as to counterbalance the further spreading of the overhead by more intense use. Finally, for given capacity, cost per unit falls continuously until full utilization is achieved. This applies both to increased route frequency spreading station overheads and to the effect of increasing load factor in spreading flying costs.

From this analysis we may draw some conclusions concerning the approximate ranges of cost for air transport. Costs escapable with respect to the individual passenger are near zero; even with respect to the flight escapable costs may be as low as 2d per seat mile. Average total cost per seat mile, however, will be at least 4d even under the most favourable conditions. The actual cost per passenger mile will then depend on the extent to which the higher capacity, lower cost aircraft can be fully utilized.

The implications for surface transport

Competition between air and surface transport is, as yet, unregulated. The Air Transport Licensing Board has been directed to exercise its functions in such a way as to further the development of British civil aviation. This has been interpreted to include the prevention of any demonstrably wasteful duplication on air routes but not to imply any concern at all for the effect of air competition on existing surface facilities.

But, in any case, how serious a competitor is air transport in the domestic market? The Beeching report argues that it is only a serious competitor on the routes from London to Manchester, Newcastle and Glasgow; it continues scathingly, 'nor is it competitive in terms of cost except while operating as the minority carrier able to keep a high load factor by creaming from the total flow'.[1] It is envisaged that improved rail services to Newcastle and Manchester will meet this challenge, while on the Scottish route a reduction in day-time rail service and concentration on the greater comfort of the overnight rail journey is anticipated. Certainly night flights from London to Glasgow could not be offered at less than 2d per mile except on the basis of variable costs only, for a very high load factor, and in conjunction with remunerative mail contracts. A regular service could only be operated profitably at fares nearly three times as high as this.[2] For instance, in 1951, when a separate

---

[1] *The Reshaping of British Railways*, op. cit., p. 13.

[2] The competition may have been underestimated in some respects. With lower capital costs some of the independent air companies may be able to compete on cost, at least for off peak traffic. For instance, in August 1963 the A.T.L.B. received an application for a service from Leeds to London from BKS air transport during winter weekends at a fare undercutting the first-class rail fare.

account was produced for the B.E.A. London-Paris service, only a slim profit was shown at fares of over 6d per passenger mile, despite extremely favourable route frequency and loading conditions.[1] This was largely due to the high incidence of landing and handling costs and the poor crew utilization resulting from the frequent turn round; all conditions very relevant to British domestic routes.

The B.E.A. experiment with even cheaper 'walk on' fares (London to Glasgow for 42/–) is an attempt to use price discrimination to improve load factors in a situation where the elasticity of demand in general is believed to be low (i.e. total revenue would be expected to decline as a result of a general reduction of fares). In so far as passenger traffic can be divided into two categories, business and pleasure, the first insensitive to price changes but requiring a guaranteed service and the latter relatively insensitive to standard of service but highly sensitive to price, the experiment could meet with considerable success. In a similar way the railway could (and probably will) use this form of discrimination on their own high quality, reserved seat services.

We have so far omitted mention of charter flying. A certain amount of surface travel is eliminated even by international charter flights. However, much of the traffic involved is highly seasonal; both road and rail may therefore benefit in so far as peak period congestion is reduced and they are relieved from the provision of capacity which would be seriously underutilized for most of the year.

We may conclude therefore that the immediate impact of air transport on surface travel is likely to be limited but that for certain trunk routes it could rapidly develop sufficiently to undermine the profitability of existing rail passenger transport facilities.

COASTAL SHIPPING

In contrast to air transport coastal shipping is almost totally concerned with freight. Where passenger traffic is important, as for instance in the Irish Sea and amongst the Scottish Islands, it is the complement of other forms of surface transport rather than a competitor. Hence it is rarely responsible for any significant transfer of passengers away from inland surface transport.

Even as a freight transporting agency coastal shipping is often dismissed as a declining or even as an already negligible element. For instance the Rochdale Committee dismissed it very curtly in the following terms, ' . . . our dry cargo coastal trade is subject to strong

[1] See P. G. Masefield, 'Some Economic Factors in Air Transport Operation', *Journal of the Institute of Transport*, Vol. 24, No. 3, March 1951.

TABLE 26

The British Dry Cargo Coasting Trade, 1948

Origin/Destination (% of Total Traffic)

Area of Unloading	Area of Loading										
	West Scotland (Clyde)	East Scotland (Forth)	North East	Wash, Thames, E. Anglia	South, S.E., and C.I.	South-West	Bristol Channel	Wales	Mersey and N.W.	Ireland and I. of M.	
West Scotland (Clyde)	·3	·1	·1	·2	0	0	·2	·1	·3	·2	1·7
East Scotland (Forth)	0	·4	1·4	1·3	0	·1	·1	0	0	0	3·4
North East	0	·1	·6	1·6	·1	0	0	—	0	0	2·5
Wash, Thames, E. Anglia	0	2·3	49·7	·9	0	·9	3·6	·4	·1	0	57·8
South, S.E. and C. I.	0	0	7·7	·2	·1	·3	·6	0	0	0	9·1
South West	0	·1	1·9	·1	0	0	1·0	—	·2	0	3·3
Bristol Channel	0	—	·2	·1	0	·1	·3	·1	·2	·1	1·1
Wales	0	—	0	—	0	0	0	0	0	0	0
Mersey and N.W.	·4	0	0	·1	0	·2	2·0	1·4	·7	·9	5·8
Ireland and I. of M.	4·4	·1	·7	·6	0	0	2·1	0	7·5	·1	15·4
	5·2	3·2	62·2	5·1	·4	1·6	9·8	2·1	9·1	1·5	100

competition from inland transport and may lose more ground, while coastwise tanker traffic, which has increased greatly since the war, is vulnerable to the development of pipelines'.[1]

In view of this pessimism we may find the share of domestic freight traffic which coastal shipping still retains surprisingly large. In 1958 it was estimated that coastal shipping carried approximately 50 million tons of traffic, or 4 per cent of the total transported. Because of the great average length of haul, however, this amounted to 20 per cent of the total ton mileage, even when deflated into inland mileage equivalents. The Rochdale Committee estimated total traffic in 1961 still to be in the region of 50 million tons so that the sector is clearly still quite significant.

The last really detailed investigation of the commodities carried and the journeys made in coastal shipping used 1948 data.[2] Such evidence as is available suggests that the pattern of dry cargo coastal trade has not changed a great deal over the years since, though there has been a very great and rapid expansion in tanker traffic. The findings of this survey are summarized very briefly in Tables 26 and 27. Particularly noteworthy is the predominance of coal traffic in Table 26 and of flow from the north-east to the south in Table 27.

TABLE 27

Chief Commodities Carried in the Dry Cargo Coasting Trade, 1948

Commodity	% of total flow	Routing
Coal	84	Tyne, Tees, Humber—Thames, S.E. (55%)
		Mersey, Bristol Channel—Ireland (10%)
Cement	2·3	London—Forth, Tyne
Sand, Stone,	4·3	Wales—Mersey
Chalk, etc.		South West—Thames
Other Minerals	2·5	
Food, Feeding		
Stuffs	3·5	

Source: Ford and Bound, *op. cit.*

There are two important consequences of this type of traffic flow. Firstly, the commodities involved are for the most part bulk goods which might be expected, when using inland transport, to travel by rail rather than by road. Even when carried by sea some ancillary haulage is required to and from the ports used. Hence the railways,

[1] Report of the Committee of Inquiry into the Major Ports of Great Britain, 1962, Cmnd. 1,842, para. 22, p. 12.

[2] P. Ford and J. A. Bound, *Coastwise Shipping and the Small Ports*, Oxford, 1951.

given freedom to discriminate in their charges, could well divert traffic from sea to rail by either charging very low rates for the trunk haul or very high rates for the short ancillary haul. In fact the coastal shipping industry complained very bitterly during the inter-war years against the cross-subsidization of coal traffic by the railways. In deference to the fears of the industry on this score clause 55 of the 1962 Transport Act enables coastal shippers to challenge rail rates on either of these grounds. The Minister undertook to refer any such complaint to an accountant of high standing, the investigation to be based on the full cost, not merely the direct cost, of the service concerned. This restriction on the Railway Board is in marked contrast to its freedom in almost all other charges. This degree of protection should ensure a modicum of traffic to the coastal shipping industry in the foreseeable future in competition with rail haulage. Indeed the Beeching report does not indicate any intention on the part of the railways to win long distance coal traffic from shipping, reporting that 'a substantial proportion of the traffic passing coastwise is carried to the ports by rail, a combination of function which is probably the most suitable and economic'.[1]

The second important consequence of the traditional pattern of British coastal shipping is the very large proportion of sailings in ballast due to the disparity of flows in opposite directions on most routes. In 1948 Ford and Bound estimated that the minimum empty sailing which would have been necessary in order to achieve the work done would have accounted for 44 per cent of the total capacity sailings. In fact in that year over 50 per cent of the capacity departing in the coastal trades was in ballast. In 1961 departures in ballast still accounted for 48 per cent of capacity sailing. A large part of coaster capacity is engaged in carrying cargo regularly, in one direction only, between ports. The tramp ship in the coastal trades is not in general the nomadic, irregular carrier we associate with the term when thinking of international trade. Nor are tramp rates the highly fluctuating results of an extremely competitive process as we might imagine them to be. Instead they are comprised of a very detailed and precise schedule negotiated through the Chamber of Shipping with the principal chartering associations.

In view of this essential regularity and stability of most of the traffics it should be possible for both ships and dock facilities to be constructed specifically for particular trades in order to meet the requirements in the most efficient manner possible. It is along these lines that the Beeching report indicates railway development should

---

[1] *The Reshaping of British Railways, op. cit.*, p. 92.

take place and the Rochdale report suggested similar development in the docks.

The economics of tramp shipping are relatively straightforward. Ships are chartered by the trip, the charterer being responsible for loading and unloading and the shipowner for the manning, victualling and operating costs of the ship. Hence costs to the shipowner can be divided into organization overheads and voyage variable costs, there being no handling costs which vary with loading. The additional cost of making a trip rather than lying idle, which constitute the lowest acceptable charter, consist of the paid out variable costs less the costs which would be incurred if the ship were to be laid up. It will thus pay tramps to remain in operation even at a loss on variable costs so long as the loss involved in operation is less than the loss incurred in laying up.

Combination amongst shipowners in negotiating the rates for coastal work cannot protect them against competition from internal transport and the Chamber of Shipping reported in 1962-3 that at the established rates for the home trades few owners can earn sufficient revenue to meet depreciation on their ships.[1] If this be the case then either more efficient operation must be achieved or coastal shipping must decline. More efficient operation might be achieved by the introduction of larger and faster ships, a trend currently to be observed, and by the more efficient terminal operations already mentioned. The evidence suggests however that, despite the fact that the Railways Board do not intend to win traffic from coasters according to the Beeching report, coastal shipping traffic is unlikely to increase and may well decrease in the near future.

Of the 50 million tons carried in the coastal trades in 1961 about 20 million was tanker traffic. Here again prospects are not bright, for though the tanker is still the cheapest form of transport per mile for liquid products the fact that so many of the major markets are inland and require a good deal of ancillary haulage after shipping puts the pipeline in an advantageous position. Hence the very rapid increase in coastal tanker traffic which has taken place over the last decade may not continue.

PIPELINES

Pipeline transportation provides a third potential challenge to road and rail transport; both the Rochdale and Beeching reports refer to a probable extension in this field. Moreover there is some likelihood

[1] Annual Report of the Chamber of Shipping of the United Kingdom, 1962-3, p. 61.

that it may extend from the traditional piped commodities, gas and oil, into other fields such as chemicals and coal. Clearly it is worthy of some investigation.

Pipelines provide a form of transport with natural characteristics quite distinct from other forms which we have been considering. There are four main differences:[1]

(i) It provides for the continuous movement of commodities in bulk at a constant rate as opposed to the transmission of variable amounts at irregular intervals.

(ii) It is geographically inflexible, being unable to respond to changes in the location of either supply of or demand for the commodity carried.

(iii) It is relatively inflexible in its ability to carry a variety of commodities. Though several petroleum products, passed in suitable sequence, may use the same pipe without any great waste through mixing and pollution, for any greater range of products, or for frequent changes of commodity, the costs of separation and cleaning appear to be significant.

(iv) It is very economical in utilization of space. Once the pipeline is laid the land above it can continue to be used for any purpose except the construction of permanent structures, subject only to the right of access for the pipeline company when repairs and replacement become necessary.

Corresponding to the distinctive physical characteristics of pipelines there is a distinctive cost structure for pipeline operation.

Firstly, the capital costs form a very high proportion of the total costs of operation; in the case of oil they constitute between 65 and 75 per cent of the total. Consequently it is most important to ensure maximum use of capacity. So important is this that in cases where there are fluctuations in demand or supply it will probably be economic to provide storage facilities either at source or destination rather than extend or duplicate the pipe facility. In the case of gas the use of natural reservoirs in suitable geological strata may make this a relatively inexpensive way of securing full utilization of the pipe, during periods of low demand. A second corollary follows from the nature of the capital costs. Both the cost of pumping stations and of the pipe itself vary fairly directly with distance. The terminal costs are relatively small. Consequently there is no very distinct taper in

[1] See G. Manners, 'The Pipeline Revolution', *Geography*, Vol. 47, Part 2, April 1962; also M. E. Hubbard, 'Pipelines in Relation to other Forms of Transport', *World Power Conference*, Madrid, 1960, Paper IIIA/8.

the cost per ton-mile as the length of haul increases. Thus the cost per ton-mile is very sensitive to the regularity of flow but not to the length of the pipe.

Secondly, there are important economies of scale in pipeline transport. As the diameter of the pipe increases its capacity grows more rapidly than the costs of laying, maintaining, and operating it. For instance it has been estimated that for the piping of crude oil the cost per barrel per mile through a 30 inch diameter pipe is only a quarter of the cost through a 10¼ inch diameter pipe when both are working at optimum capacity.[1] Therefore, to obtain really low cost piping, the flow has not only to be regular but also to be large.

Thirdly, the cost or piping varies between commodities. The greater the viscosity of the product to be carried the more difficult it is to pump and therefore the greater the cost. For instance, crude oil is far more expensive to pipe over long distances than light petroleum products because of the need to heat it to increase its liquidity. For a commodity to be suitable for piping it must be either naturally in a highly liquid form, easily transformable into this form or able to be suspended in some highly liquid carrier. Thus chemicals carried in solution or minerals carried in suspension are also capable of pipe-line transportation; whether it is an economic proposition will depend on the cost of the transformation into liquid form and the subsequent reconversion into a usable form.[2]

The crucial question is how these costs of transportation compare with the costs of other agencies. So far there is little British experience on which to judge but estimates of the costs of transporting oil by alternative methods in Europe suggest that even a pipe with a capacity of one million tons per annum would be considerably cheaper than road or rail.[3]

Britain has not yet developed a commercial pipeline network. 1,200 miles of pipeline were laid during the war to supply military airfields and though this network has been used a little for commercial purposes recently, it is not a very convenient network. Several shorter pipelines have been constructed since the war by the

[1] L. Cookenboo, Jnr., 'Economies of Scale in the Operation of Crude Oil Pipelines'. Unpublished Ph.D. Thesis, Massachusetts Institute of Technology, 1953, p. 188 (referred to in Manners, *op. cit.*).

[2] A mixture of crushed coal and water is pumped well over 100 miles for power station use at Cleveland, Ohio. The development of a 'stabilized slurry' with a high coal content may well permit both easier piping and direct use as a liquid fuel, thus halving processing costs (see Manners, *op. cit.*).

[3] M. E. Hubbard, *op. cit.*

[4] See H. G. B. Nolan, 'Oil Pipelines in the U.K.', *The Petroleum Times*, June 30, July 14, July 28, 1961.

MAJOR OIL & GAS
PIPELINE PROPOSALS

———— Oil, already constructed
—G— Oil, Government Pipeline
········ U.K. Oil Pipelines Project
- - - - Proposed Methane
Distribution Grid

A   Shell Haven
        Coryton
        Thames Estuary
        Oil Storages

B   London Airport

Finnart
Grangemouth

Leeds
Manchester
Stanlow   Partington   Sheffield
                             Nottingham
                    Peterborough
Birmingham
Worcester   Hertford
Milford   Gloucester
Haven
Llandarcy   Severnside   Reading   A
                    Aldermaston   B
                Fawley

Figure 12

major oil companies,[4] either to carry crude oil from the ocean terminal to an inland refinery (Tranmere–Stanlow, Milford Haven–Llandarcy, Finnart–Grangemouth) or to carry products of various kinds from the refinery to a market (Fawley–London, Fawley–Avonmouth, Stanlow–Partington, Isle of Grain–London) (see Fig. 12).

The earlier projects, approved by private acts of Parliament, aroused little public interest. Opposition to the Thames Estuary–Midlands–Merseyside pipeline and the projected Esso line from Fawley to London Airport provoked some controversy as a result of which the Government have, by the Pipe-Lines Act, 1962, assumed considerable powers of direction over future construction. The Act makes it necessary to secure ministerial permission for the construction of any major pipeline.[1] In exercising his discretion the Minister of Power is to prevent the construction of superfluous pipelines, to which end he can impose common carrier obligations on the pipeline owner at rates regulated by himself. In this way it is hoped that the economies of scale may be achieved in the construction of the line and that it may also be operated at high capacity in order to keep down costs. The Minister may also make orders for the compulsory acquisition of rights over land required for the construction of pipelines, fair terms of compensation, in the default of agreement, being decided by the Lands Tribunal. Additional powers to control the mode of construction, safety provisions, examination and repair make pipeline transport as heavily subject to public supervision as the railways in the late nineteenth century. By July 1963 three applications had been made for authorization to construct pipelines. That of the Rugby Portland Cement Company for a 10 inch diameter pipe to convey annually about one million tons of chalk, slurried in water, from a quarry near Dunstable to a cement works at Rugby was accepted in May.

The other two concerned competing projects for the conveyance of lights oils. In mid-July the Minister announced the acceptance of the U.K. Oil Pipelines Ltd. application and the rejection of the rival scheme of Trunk Pipelines Ltd. on the grounds that the former would cost less both to construct and to operate.[2] The statutory procedure still provides for a public inquiry, if necessary, to deal with objec-

---

[1] Defined as any line more than 10 miles in length.
[2] U.K. Oil Pipelines Ltd. is a consortium of five major oil companies. Their 317-mile project will link London and the Midlands with the refineries of the Thames Estuary, Southampton Water and the Mersey by a direct route (see Fig. 11). Trunk Pipelines Ltd., a company formed by a group of engineers and merchant bankers, aimed to link the same areas but by a more devious route, using existing easements of British Railways, British Waterways, the North Thames Gas Board and L.C.C. sewers.

tions and not until after such objections have been considered is permission to construct finalized.

At present about 5 million tons of oil fuel is transported by rail; this represents the maximum possible loss of rail traffic. Road losses may not be so great because road haulage of oil products is mostly concerned with local distribution rather than with the trunk haul from refinery to distribution depot. Coastal shipping and inland waterways may lose considerable traffic despite the fact that ships can compete favourably, mile for mile, with all but the largest pipes, as the cross country pipe permits a much shorter haul between points than waterborne transport.

For other commodities the future of pipeline transport in this country is not so assured. Where sufficiently large and regular flows of minerals are available to support a pipeline it appears that continuously coupled railway trains, working a shuttle service, may prove more economic, especially when the costs of processing for piping are taken into consideration. The agreement already reached between the Railways Board, the N.C.B. and C.E.G.B. envisages considerable initial expenditure on new equipment to facilitate such a service in the confident expectation that costs of coal transport will be considerably reduced by the investment.

There is one further pipeline development which may have some effect on road and rail transport, namely trunk piping of gas. This would eliminate the necessity to transport coal over long distances due to the gasification of coal *in situ* and the import and piping of Saharan methane gas. There is also currently under construction a line to supply liquid petroleum gases to London from the Esso Fawley refinery. In 1961 rail carryings of coal to gas works amounted to 11 million tons. Though most of this was relatively short haul traffic it is nevertheless potentially very profitable by virtue of its loading characteristics and the regularity of the flow; its loss would be a serious blow to the railways.

It appears likely therefore that the development of trunk pipelines for petroleum products and for gas could reduce the demand for rail and sea transport of oil and coal. Particularly in the case of petroleum, however, it would not affect road transport so seriously as most of the product tankers on the roads are involved in the later stages of distribution and not in trunk haulage.

CONCLUSIONS

The effects of competition from other transport agencies for road and rail appear at present to be only marginal. However, both air and pipeline transport could well develop rapidly in the near future to the exclusion of road or rail transport in certain sectors. It is desirable that any such development should be orderly, not involving wasteful duplication of facilities, yet at the same time that any control imposed should not prejudice the chances of further technological developments in these agencies.

PART IV

CO-ORDINATION THROUGH THE MARKET

CHAPTER TWELVE

The Alternative Means of Co-ordination

The British transport scene that we have described is basically competitive. Road, rail, air and waterborne transport vie for the traffic forthcoming. No single agency is able to operate its capital equipment at full capacity due to competition from the others. Would it not therefore be better to concentrate all traffic on particular routes or of particular types into an individual agency, and hence avoid the waste involved in operation under conditions of excess capacity? Might we not profitably emulate the planned transport system of the U.S.S.R. where the bulk of freight transport is concentrated on the railways?[1]

Two answers to this question have already been made explicit.

(i) Traders may be interested not only in the cost of transport but also in the qualitative aspects of the service. A planned concentration on one agency involving use of capacity at a high load factor may detrimentally affect the quality of service provided. In the absence of any alternatives the trader will be unable to express any special valuation which he would place on refinements of service.

(ii) Moreover, such a planned solution may tend to be inflexible. The fact that it may currently reduce cost to concentrate on rail transport does not mean that it need necessarily always do so. If the development of alternative forms of transport is discouraged, desirable technological opportunities may be overlooked in the future.

The objection to a physically planned concentration amounts to the judgment that such a solution is inimical to (a) choice, and (b) development in the transport sphere. For instance, a recent commentator on the Soviet system has noted that, 'For years it was impossible, without the necessary "blat" (the Russian word for spivvery), to get a railway ticket from Kiev to Khartov.'[2] During peak periods people might wait for two days at a time at a station trying

---

[1] For a description of the essential elements of Russian freight transport see Ernest W. Williams, Jr., *Freight Transportation in the Soviet Union*; N.B.E.R. Research Study, No. 76, 1962.

[2] E. Crankshaw, *Krushchev's Russia*, Baltimore, 1959, p. 73.

to board a train. Clearly this excess demand could have been stifled by an increase in the price of passenger travel; the failure to take such a step indicates the unwillingness of the authorities to permit the consumer to express the insistence of his demand in monetary terms. The simultaneous reluctance to expand capacity to meet demand at existing price levels indicates the view of authorities that this particular demand was 'unnecessary'. Hence the absence of choice or freedom of development in Soviet passenger transport represents the imposition of an authoritarian scale of priorities which differs from the scale which would result from free consumer choice under the existing income distribution. It has been argued that such an imposition may be necessary because the individual cannot recognize the full implications of his choice. But charging a price equal to the full long run social marginal cost would enforce such awareness. Hence a planning solution, as opposed to use of price mechanism to direct activities, may constitute an attempt to deny the right, to individuals or to traders, to compare transport as an alternative to all other goods consumed or factors used in production.

There is one respect, however, in which a planning solution may be the only way to avoid a mistaken outcome from a competitive situation. W. A. Lewis considers[1] the case of competition between two agencies for four traffics, *a, b, c, d*. Each agency is assumed to have a given level of indivisible cost in the short run but on divisible costs one agency has the advantage for two of the traffics and the second agency has lower divisible costs for the other two traffics. It could transpire that even by perfect discrimination (i.e. charging a price only just less than divisible cost of the competitor) it would still not be possible for both agencies to operate profitably or even for one to operate profitably whilst the other agency obtained that traffic for which it had the lower direct costs. Consider the following example:

Indivisible cost		Divisible cost				Total
		Traffic a	b	c	d	
Agency X	4	1	1	3	3	12
Agency Y	6	2	3	1	1	13

Neither agency can pay its way if both services are offered in competition. X will secure traffics 'a' and 'b' from which it will get a maximum revenue of 5 while it takes 6 to keep alive: Y will secure traffics 'c' and 'd' from which it will draw a maximum revenue of 6 while it needs 8 to keep alive. The total cost if the two industries were operated would be 14 and neither would be profitable, whereas if only one agency were to operate then the cost would be either 12 (for X)

[1] W. A. Lewis, *Overhead Costs*, Chapter 1.

or 13 (for Y). In a situation of this kind, where the overhead costs are not divisible, it might be sensible from the outset to decide which agency would give the lowest total cost for the service to be provided and choose that agency as a monopoly operator, unless there is evidence to suggest that demand might increase sufficiently to support both in the future.

If both agencies are working at minimal indivisible capacity for this route, yet within this capacity could carry all four traffics, the best outcome would be for one of the agencies to cease to compete in the long run (i.e. not renew its capacity). If both sets of equipment came up for renewal simultaneously then it would be best if that agency with the probability of higher total cost in the future, when carrying all four traffics, withdrew. If not the best outcome would depend on the complex relationship of the lag between the necessary renewal times, the difference in potential cost levels and the rate of the discount used in the comparison. But there are several reasons why the desirable outcome might not ensue.

(i) Firstly, it may not be the case that the agencies are currently discriminating equally well in pricing. In such a situation the less rigorous discriminator would tend to make larger losses and hence be inclined to drop out first.

(ii) Secondly, the preferred operator (assuming the relevant rate of time discount to be known) might meet the decision concerning the incurring of a further block of indivisible cost first. If he can see no prospect of his competitor withdrawing, he may himself withdraw in order to avoid losses. Under the conditions outlined above this would be an improvement on the current position (assuming that the ensuing monopoly situation did not itself induce a rise in costs and prices), though it would not be the best available outcome.

(iii) Thirdly, undue optimism on behalf of both of the agencies may cause them to continue to compete, thinking that a profitable outcome will eventually be possible in a competitive situation. Such an outcome would depend on either an expansion of traffic offered or a reduction of direct costs for one of the agencies in order to attract traffic from the other. If neither of these developments is likely, a planned abolition of duplication should be undertaken.

This case may be relevant in several spheres of British transport. In the provision of transport facilities to rural areas, in the competition between rail and air on some of the major trunk routes, and also in the prospect of competition between air operators on some of the established air routes, we can discern the danger of duplication of facilities involving waste capacity far in excess of that required as the guarantee of high quality service. In the first and third of these cases the ultimate controlling decision lies in the hands of the relevant ministers who in both cases appear to be aware of the nature of the problem. In the second case, however, partly because two different ministries are involved and partly because the routes concerned look potentially profitable to both operators, wasteful duplication may continue.

We have argued that the problem of co-ordination is to ensure that the required transport services are provided with the minimum call on the real resources of the nation. Though this would appear to be the sort of task ideally within the competence of a central transport co-ordinating body, responsible for the operation of all forms of transport and allocating traffic between the different agencies, we have criticized such a solution on the grounds that such a central body cannot adequately assess the value to the user of the many and varied aspects of a transport service beyond the mere transfer of a consignment from one locality to another. We can only feel sure that the user is obtaining the best possible service if he is able to choose for himself between various forms of transport. We must notice, however, that, especially where competing agencies are operating at less than full capacity, this freedom of choice itself involves cost; a resolution to permit freedom of choice amounts to the judgment that in the long run the superior flexibility and adaptability of a system involving choice outweighs the extra cost of capacity in the short run implicit in such a situation. Such a judgment is not necessarily of universal validity and there may be situations or occasions where capital is deemed particularly scarce and which would therefore justify central direction of traffic in order to minimize capital requirements. The early stages of development of a country may provide exactly such conditions.

If, however, we do accept the need for the consumer to indicate, by his choice of agency, the exact transport product-mix which he considers best suited to his requirements, then the task of public policy is to ensure a structure of relative charges which reflects the relative costs of the alternatives provided. Or, to put it another way, the task of the co-ordinator is to ensure the comparability of the prices charged by different agencies. Hitherto we have described the current state of affairs in British transport and analysed the way in which, and the extent to which, this has been satisfactorily achieved. Several defects are apparent and we have commented on these already. But we now have to ask the question whether, within the framework of the trader's freedom to choose his transport agency, there is any radically different approach which could have been, or could still be, taken to achieve effective co-ordination.

There are three such alternatives which are worthy of consideration.

THE MANCE SOLUTION

The first of these solutions was advocated by Brigadier-General Sir H. Osborne Mance in his book *The Road and Rail Transport Problem* published in 1940 and subsequently in papers read to the Institute of Transport in 1941 and 1951.[1]

The crux of the problem according to this viewpoint lies in the fact that, whereas railway operators are responsible for their own track, the road operator has his track provided for him by the Government and is charged a price in the form of licence duty and fuel taxation, the incidence of which is almost purely random. Mance argues that, as a result, competition tends to be pursued on the basis of something roughly approaching expected average costs rather than on the desirable basis of out of pocket expenses. This could be avoided by making the provision of track for all forms of transport a Government responsibility and allowing competition to be waged on the grounds of the out of pocket expenses of providing a given service by various agencies. Then, he argues, we should obtain an optimum use of capacity. 'If it is possible,' he says, 'for all transport to contribute to the fixed costs of transport as a whole on an identical basis, the competition on the basis of out of pocket expenses would be on a footing of equality never yet attained.'[2]

Even assuming that an entirely satisfactory distinction can be made between Mance's two categories of track costs and out of pocket expenses, the suggestion has obvious defects. A competitive solution based on the relative running costs of two agencies will tend to favour the agency with the higher overheads. If track costs had already been incurred, and there was no question of any replacement or further investment being necessary in this respect, all would be well. With the necessity for both replacement and new investment in track (both road and rail) over time, the solution fails to take into account an important real cost to the community in deciding how

[1] Brigadier-General Sir H. O. Mance, 'The Road and Rail Transport Problem', *Journal of the Institute of Transport*, Vol. 21, No. 8, July 1941; also 'The Perennial Question of Co-ordination', *Journal of the Institute of Transport*, Vol. 24, No. 2, January 1951.

[2] *Journal of the Institute of Transport*, July 1941, *op. cit.*, p. 252.

traffic is to be allocated. Moreover, the prescription that the toll for all track be levied from public and private road hauliers and from the railways on an identical basis (possibly the ton-mile or passenger-mile) implies cross-subsidy not only of the agency with higher track costs by that with lower track costs, but also cross-subsidy within each agency as the costs involved by the occupation of track are not always the same within either road or rail transport. Hence, even in purely static terms, this is not a satisfactory solution. There will, furthermore, be immense complications when we try to devise criteria for investment in track in these circumstances.

THE SARGENT SOLUTION

Mr J. R. Sargent has more recently suggested[1] an alternative solution to the problem of meeting the indirect (mostly track) costs of road and rail transport. He argues that both road and rail should be responsible for meeting their own indirect costs so that there should be no cross-subsidization between the agencies. In this respect he disagrees with Mance.

He does, however, argue that both in road and rail, indirect costs should be met through an even distribution (per mile run) of the total burden, thus following Mance in this respect. Thus the railways should be required to fix the charge for every service so that it exceeds the direct cost of the service by a uniform amount. Otherwise, Mr Sargent argues, if the railways are given complete commercial freedom they will use their power to discriminate in such a way as to absolve from the burden of contributing to overheads all traffic for which there is serious road competition. As the road haulier must pay petrol tax irrespective of the cargo carried he will have to include this provision to meet track cost even in the case of the most marginal traffic. Hence there will be a bias in favour of fuller utilization of railway than of road capacity.

In so far as the petrol tax is in excess of the full marginal social cost of track use in road haulage (including wear and tear and congestion costs), Mr Sargent is correct in arguing that divergent ways of meeting fixed track costs constitute an incomparability between the charges which competing agencies can levy. But in arguing that local exceptions to the general rule against discrimination might be allowed on a limited proportion of the traffic, when this would make possible the fuller use of capacity which might not otherwise be used, he is expounding the deficiencies of his own case. The whole purpose of using discriminatory charging methods in rail freight transport is

[1] J. R. Sargent, *British Transport Policy*, Oxford, 1958.

to secure fuller utilization of existing capacity. In doing this the burden of overheads can be spread and the average charge reduced. Hence discrimination does not necessarily mean a cross-subsidy within rail traffic in any useful economic sense so long as no traffic is accepted at a rate below the marginal cost of transporting it.

But what about the comparison between road and rail rates? If rail transport is able to charge short run marginal cost and road must charge short run marginal cost *plus* the petrol tax contribution to track cost, does not this automatically imply that traffic will be attracted to the railways even where the real opportunity cost, in terms of call on the communities' resources, is less by road? If the petrol tax were always in excess of the marginal social cost of track use Mr Sargent would be right. But, where track occupation causes congestion and requires new road construction it may well be the case that the petrol tax is too small and that traffic is attracted to use this part of the road network though the alternative rail facility would make less call on national resources.

We must agree with Mr Sargent's diagnosis that there is some distortion in the present state of affairs though we would disagree with his contention that this must always be in favour of the railways.

The Sargent remedy, however, is not acceptable. To allocate total indirect costs in proportion to the direct costs involved in both road and rail transport is to eliminate any incentive to intensive use of existing capacity and, in the case of rail transport at least, would discourage new investment in any asset which required intensive use to render it profitable. In railway operation it is possible to allocate a large proportion of indirect cost to particular types of traffic or to particular sections of operation. The ratio of direct to indirect costs is not the same for all types of working or areas of operation. To require the railways to average out these costs is to take a retrograde step away from the identification of costs and the linking of price to opportunity cost. The outcome of such a policy would be to discourage the railways from developing and investing in those forms of bulk transportation to which they are most likely to be suited and which could be made profitable by use of the incentive of a low price, including a relatively small proportionate contribution to overheads.

A great deal has been made in the past of two other arguments. The railways claim that road hauliers are favoured by the fact that they are not required to pay interest on capital previously invested in roads whilst the railways are (though the 1962 Transport Act has removed some of the substance of this argument). The road users on the other hand point out that the taxation system discriminates against them not only by making their contribution to overheads fall as a direct cost but also by the fact that this annual 'contribution' amounts to more than twice the annual expenditure on roads. Under Mr Sargent's régime the outcome of these arguments must have an effect on the price of every service provided so that the relative price structure and hence the allocation of traffic would depend on what could only be arbitrary answers to questions which are not uniquely soluble even in principle.

The argument for meeting indirect costs by discrimination is that these costs, once incurred, are inescapable so that the problem of securing an efficient use of resources then resolves itself into one of securing that the costs incurred *in addition* to these prior costs are minimized. And this can only be ensured by making the comparison between two agencies rest on the variable costs involved. Discrimination, by only adding to these direct costs an amount 'that the traffic can bear', will not affect this allocation of traffic whilst permitting prior commitments to be met.

Moreover such an argument does not depend on the size of unit under consideration. It can be applied equally to the problem of what minimum charge is economic for a fill-in load as to the problem of whether to provide a particular full train service or retain a branch line in operation. It requires careful thought, detailed costings and a good deal of expertise in estimation of demand, but these are attributes which are required of any advanced industrial management.

THE DIFFERENTIAL ROAD TAX SOLUTION

Both Mance and Sargent started by affirming their desire to seek a solution to the problem of co-ordination through the price mechanism by relating price to those costs which they took to be relevant. In both cases their suggestions consisted of making rail pricing policy comparable with that currently applied in road transport, the latter being taken as datum. The most recent contribution to this debate has been the observation that co-ordination through the price mechanism requires charges to be related more precisely to social cost for the use of roads as much as for other assets.

The theory of efficient motor taxes began with Pigou[1] who argued that taxation should attempt to eliminate the difference between marginal private cost and marginal social cost. If we assume that cost is inversely proportional to speed and that beyond a certain traffic density each additional vehicle affects the performance of other vehicles on the road then marginal social cost (m.s.c.) exceeds marginal private cost (m.p.c.) by a discernible amount, i.e. m.s.c. =

[1] A. C. Pigou, *Wealth and Welfare*, 1st edition, 1920.

m.p.c. $(1+\mu$ m.p.c.$)$, where μ m.p.c. is the elasticity of marginal private cost with respect to amount of traffic.

More recently, A. A. Walters has made some estimates of these discrepancies both with U.K. and U.S. data. Using a traffic survey carried out on Slough High Street in 1952 he examines the various factors which affect traffic speed. Concentrating on the effect of traffic density he estimated that the optimum tax for use of this stretch of road should be somewhere between 3½d and 9d per vehicle mile according to time of day and traffic conditions instead of the 1·1 to 1·6d per mile which was at the time being paid through the petrol taxation system.[1] He also argues for much higher parking charges in congested areas (over 2s per vehicle hour would have been efficient for Slough High Street).[2]

Pigou and Walters have thus suggested that the function of an efficient charging system for road use is to make each road user aware of, and directly responsible for, the extra cost which the rest of the community is forced to incur as the result of his decision to use the road at the particular time and place that he chooses. But is this all that should be required? It has recently been suggested that there are four conditions which might sensibly be required of any method of approaching the problem of controlled road use.[3]

(a) It should result in a usage of existing roads as close as possible to the theoretical optimum where the benefits to the marginal user were just sufficient to meet the marginal social cost resulting from this unit of road use.

(b) The cost of administering the control system should be less than the advantages accruing from optimum use as compared with the sub-optimum alternative available without the control system.

(c) It should provide guidance to a rational road investment policy by showing where demand for roads is greatest.

(d) It must appear fair and acceptable to the public.

To charge a price equal to marginal social cost would appear to be acceptable on grounds (a), at least within the bounds of possibility, on ground (b) if a serious attempt were made to solve the engineering

[1] A. A. Walters, 'Optimum Motor Taxes', *Applied Statistics*, Vol. X, No. 3, November 1961. An analysis of American data yields a similar conclusion that marginal rates of road taxation are too low in urban areas.

[2] A. A. Walters, 'The Theory and Measurement of Private and Social Cost of Highway Congestion', *Econometrica*, Vol. 29, No. 4, October 1961.

[3] M. E. Beesley and G. J. Roth, 'Restraint of Traffic in Congested Areas', *Town Planning Review*, Vol. XXXIII, No. 3, October 1962.

problems involved,[1] and feasible even on ground (d) if it was demonstrated that the method concerned only the distribution and not the absolute level of the tax burden. The real problem arises when we attempt to interpret the results of pricing according to marginal social cost as a guide to investment. Walters points out[2] that where competing rail charges are above marginal social cost this method of pricing would still be biased in favour of more intensive use of road capacity. This would make new road investment appear more desirable than it ought to if the real cost of both forms of transport were considered. Only if the railways were able to discriminate perfectly, by charging as little as short run marginal cost where necessary in order to obtain traffic and charging what the traffic will bear on all other consignments, would traffic be sensibly allocated between agencies.

Assuming that we can determine the marginal social cost of road use and devise a method of charging a price equal to this have we a satisfactory investment criterion? D. L. Munby argues that we have. 'If we knew what people would be prepared to pay for the opportunity to travel at higher speeds, as well as the other operational costs of delays, we would know when to extend the existing road network, either by adding a new lane or building a new road. Ideally we would compare the cost of these improvements with what those excluded from the existing road would be ready to pay in order to travel, plus what people would be ready to pay to travel faster than they can on the present road. A congested road may be making a surplus over its cost of construction, etc. We would continue improving roads until such a surplus disappeared. There may be no such surplus, where the cost of land is very high, even with a high charge to limit traffic.'[3]

The road charges
The market mechanism applied to road use in the following way might meet the four requirements suggested by Beesley and Roth. Road charges would consist of three parts.

(a) *User charge.* The road system would be split into sections which would be fairly homogeneous for pricing purposes, e.g. local urban area, rural area, inter-urban motorway, etc. On each road or section a direct charge should be levied according to the marginal social

[1] C. D. Foster, 'Electronics and Automation of the Road', *Road International*, Summer 1962.
[2] A. A. Walters, *Econometrica*, October 1961, *op. cit.*
[3] D. L. Munby, 'Roads as Economic Assets', *Bulletin of the Oxford Institute of Statistics*, Vol. 22, No. 4, November 1960.

cost of using the facility. If this were to be done electronically then it might be possible to devise a system of differential charges for road use very delicately adjusted to cost.

(b) *The overhead charge.* Secondly each area should raise by levy a sufficient amount to cover the difference between the total costs incurred in that road area, including administration expenses and capital charges, and the revenue obtained from the user charge. This might be raised by licences to run in particular areas for particular specified periods of time. The amount which would need to be raised in this way would vary according to the current state of congestion in the area. The greater the congestion, the greater the revenue from user charges and hence the lower the overhead charge need be. As new investment took place to reduce congestion the user charge would be diminished and the overhead charge raised, thus rearranging revenue in such a way as to stimulate the use of uncongested rather than congested road capacity at the margin. In some cases the overhead charge might be very low, or in principle, even negative. In such cases the surplus of user charge revenue over total cost might be returned to users in proportion to their expenditure, or retained to finance the investment which would presumably soon be required in such a situation.

(c) *Central taxation.* If the Government wished to obtain contribution from road users towards central funds then a national levy or licence fee should be used so as to minimize the distortion of either the investment decision or the decision as to the current use of roads. This would also have the additional advantage of showing exactly how much was being raised towards non-road funds from road users as such.

The investment decision
The investment decision would be taken on the basis of the surplus of user charge revenue over total cost (including both variable costs and the allocation of capital charges) on the road which it was proposed to improve or supplement. Where this surplus was sufficient to promise a satisfactory return new investment would be authorized. But this criterion used alone would mean that capacity would only be increased in response to, and in arrear of, changed traffic demand. It would therefore appear desirable that the investment decision should also take account of traffic expansion by estimating the surplus that would be forthcoming if no new investment were made. For instance if a new town were to be built in the middle of a previously

rural area it might be sensible to build a new road *in anticipation* of the traffic expected rather than to wait for a congestion surplus to accrue.

Such is the theory. Then remains the practical problem of ensuring that such a solution is technically feasible and politically acceptable. This can only be ascertained by experiment and experience. Charges for on-street parking have given a start in the right direction, it remains to be seen whether in the interest of true economy of resources this can be pursued to its logical conclusion.

CHAPTER THIRTEEN

The Prospect for the 'Sixties

In the first part of the book we insisted that the transport problem could only be dealt with satisfactorily on the basis of a 'comprehensive' view of the whole field.

The only explicit attempt to take such a view was contained in the 1947 Transport Act, and we argued that this attempt failed, due to the fact that it did not contain any mechanism or directive to ensure that traffic was allocated to that means of transport which could provide the required service most cheaply in real terms. Traders were left with the choice between road and rail transport whilst charges were not similarly related to cost in the two spheres. Misallocation of traffic was bound to result. Unfortunately, this consequence was obscured by the fact that many exponents of the 1947 approach took the view that transport should be provided as a social service with the price to the transport user not necessarily related to the cost of provision. Followers of this school tended mistakenly to regard existing means as sacrosanct, not recognizing that, even in providing a public service, there is no economic justification for supporting the less efficient of alternative modes of provision.

The demolition of the transport monopoly, which began in 1953, did not stem entirely from an enlightened recognition of this defect. Instead it was partly based on a dogmatic devotion to private enterprise by those to whom nationalization itself was the supreme insult. This attitude was equally misguided. Completely unregulated competition between agencies is no more of a solution than its converse if the private cost structures of the competing agencies are not similarly related to the real costs of operation.

It would be unfair to a series of Transport Ministers to suggest that they were unaware of the necessity to put the railways on to a satisfactory footing in order to compete with road transport, both private and public. The abolition of the obligations relating to equality and to undue preference by the 1953 Act and the large capital expenditure involved in the modernization plan are evidence of this. They did not appear to appreciate fully, however, the extent of the changes in attitude required if the railways were to be enabled to obtain that traffic for which they were best suited and to lose that which they could not carry economically. The control of the

Transport Tribunal over maximum passenger fares meant that cross-subsidization could in many cases only be withdrawn by closures and Sir Brian Robertson might well have felt loth to close lines which could conceivably have been made to pay their way if there had been complete freedom of charges. Fares remained too low on some routes and too high on others; profitable traffic was lost and unprofitable traffic retained.

Mounting B.T.C. losses, together with increasing road congestion, underlined the need for a 'new look' for the whole transport industry. From this increased interest in transport emerged four Acts of Parliament which together comprise the statutory framework within which the system will operate in the 1960s. The Acts (in chronological sequence) are:

(1) Road Traffic Act, 1960.
(2) Civil Aviation (Licensing) Act, 1960.
(3) Transport Act, 1962.
(4) Pipeline Act, 1962.

ROAD TRAFFIC ACT, 1960

The first of these measures was primarily a consolidating act, bringing together, with only minor modifications, the existing general provisions relating to road traffic and the road transport industry. The regulations for both road haulage and road passenger transport remain substantially unaltered. Even the most anomalous regulations remain.

In road haulage the grant of 'C' licences continues unrestricted whereas the 'A' or 'B' licence applicant has to demonstrate a need for his service (though the onus of proof lies on objectors).[1]

Similarly in road passenger transport existing operators are protected against potential competition, ostensibly in the interests of cross-subsidization. But the withdrawal of some road services has shown that this method is no longer equal to the burden cast upon it. The Jack Committee therefore reported in favour of support for unremunerative services administered through the county councils.[2] The Minister promised some action on these recommendations during the autumn of 1963, though even before this the Traffic Commissioners began to permit operators to make their own attempts to

---

[1] The licensing system is now being reviewed by the Geddes Committee.

[2] Similarly both the Highland Transport Inquiry and the Council for Wales and Monmouthshire recommended support for rural bus services, though in both cases they wished it to be Exchequer financed.

deal with the problem of rural services. For instance, Midland Red were, early in 1963, permitted to introduce different fares scales for urban and rural routes which they were refused on a previous application while the Jack Committee report was still pending.[1] It would be no more than a logical extension of this change of attitude to admit fares and costs as relevant evidence in all applications so that price competition might be taken into account in the granting of licences.

This might also ease the task of the Traffic Commissioners with respect to the authorization of 'reasonable fares', for if a proposed fare was unduly high a competitive 'bid' for a licence would be possible.[2] It would also enable very small local operators, able to provide a limited service at low cost, to secure licences. For instance, a village operator might economically run a service into the nearby town and back again in the evening by combining his bus operation with other business in the town.

But what about the long distance road services? These services have achieved high load factors, and therefore profitable operation, for two reasons. Firstly, restriction of entry has enabled capacity to be tailored precisely to demand. Secondly, trunk rail fares have, in the past, been above cost for this type of route as a consequence of the policy of maintaining standard mileage charges. Complete freedom of entry into the field of coach operation would certainly ensure that any necessary extensions of coach services would be forthcoming and would eliminate any monopoly profit of existing operators. It would also have the effect, however, of reducing load factors and thus eliminating the natural advantage of road over rail in relating capacity to demand. With a single rail operator, able, under the 1962 Transport Act, to make the best of whatever natural advantages that form of transport possesses, competitive comparability would seem to be best served by some continued restriction of entry in long distance coach work. Any signs of unreasonable profitability resulting

[1] Some companies already operated differential scales. Maidstone and District, for instance, have five scales, that for rural services being the most expensive. The differentials have been reduced rather than increased during the last decade, however.

[2] This already occurs, in effect, in the case of some assisted travel services where an employer makes contract arrangements but his employees pay something towards the cost of the service. Where a licence application for a service of this type arises the possession of a contract, frequently granted after competitive tendering, has been accepted by the Traffic Commissioners as important evidence. Hence the operator willing to provide a service at least cost will usually obtain the licence, so long as he does not abstract from other established services or contravene fair wage regulations. See 'Controlling Assisted Travel Services' *Bus and Coach*, February 1963, p. 65.

P

could still be dealt with by the Traffic Commissioners under their normal powers to control fares.

THE CIVIL AVIATION (LICENSING) ACT, 1960

Prior to this Act, B.E.A. monopolized the major air routes in the U.K. Independent operators could only provide charter services or operate as associates of the nationalized airline on routes which B.E.A. did not wish to operate itself. The Act provides the means whereby this monopoly position may be challenged by private operators. Because of the vast amounts of capital equipment required and the need for careful supervision on the grounds of safety, competition cannot be completely unrestricted however. The Air Transport Licensing Board, set up by the Act in place of the previous Advisory Council, is required to issue licences in such a way as to further the development of British civil aviation.[1] Though by Section 2 of the Act A.T.L.B. was enjoined to consider the adequacy of existing services, the danger of wasteful duplication, the existing or potential need or demand for any proposed service and any capital expenditure reasonably incurred by existing operators, the exact way in which such considerations should be interpreted was not set out. Thus the Board would have to create a 'case-law' which, because of the rights of appeal, would be effectively directed by the Minister. Indeed, the reversal of the first major decision of the Board, to deny Cunard Eagle a licence to fly the North Atlantic, demonstrated the reality of this ministerial control.

The future pattern of internal air services has not yet been settled. It will depend, however, on the results of two conflicts currently being carried on within the licensing system.

Firstly there is the question of competition between B.E.A. and the independents. A.T.L.B. has declared that it is willing to licence new operators on B.E.A. routes, but only to such an extent that the existing level of traffic is not affected. Hence the independents are only to be licensed in so far as they can generate new traffic or cater for the secular expansion in demand for air transport more satisfactorily than B.E.A. Ministerial modification of this approach has resulted in independents being allowed on domestic routes so long as they do not take more than one half of B.E.A.'s anticipated traffic growth. In the case of the London–Manchester route the Minister declared in September 1962, 'it would be premature, having regard to

[1] There are four types of licence; 'A' for scheduled operation, 'B' for regular chartered flights available only to specified travel agents, 'C' for occasional charters and 'D' for regular charters not available to the general public.

B.E.A.'s expenditure in developing the route, to licence a second operator at present'. This ruling was given against the recommendation of the appeals commissioner and despite the fact that the proposed competition only amounted to one fifth of B.E.A. anticipated growth. In contrast independent competition has been licensed on the London–Glasgow route.

The second of the licensing conflicts is that between the airlines and the railways. A.T.L.B. have explicitly stated that they do not consider diversion of traffic from surface carriers to be a relevant objection to the grant of an air licence.[1] However, in granting the B.E.A. application to operate winter night flights between London and Glasgow at a rate of below the normal second class rail fare the Board supported its case by arguing that, as only a limited capacity was concerned, diversion would be negligible. Which of these apparently conflicting attitudes will emerge as the rule is yet to be seen, but the refusal of competition on the London–Manchester route may also be viewed as an indication that the Minister does not wish to draw attention to any diversion of traffic from the railways.

Is there anything unobjectionable which can be said about air competition in internal transport? Certainly, we can observe that any service provided for a revenue below the marginal cost of providing it is both unprofitable to the operator and wasteful of the community's resources (unless it is in competition with an even more heavily subsidized service). The test case here would appear to be the newly licensed 2d a mile night flight between London and Glasgow.

Even on the most optimistic assumptions concerning operating costs it appears that the flight could only cover full variable cost by achieving a very high passenger load factor as well as substantial mail contracts. Failing this the service could still be justified commercially if it were held that it achieves a break-through in 'air consciousness', whereby the sales and profits of other services are enhanced. The 'true' marginal revenue would then be greater than the direct revenue from the flights.

But we may question whether this is a sufficient justification for the service. In 1961–2 B.E.A. stated that, after the apportionment of aircraft costs and the allocation of overheads, losses were sustained on all domestic services. If British Railways are to be required to break even by whatever means are necessary, but the airlines are able to operate at an overall loss, the method of discriminatory pricing will not necessarily result in the desirable utilization of capital equipment. If a service does not at least meet the costs which can, in the long run,

[1] Report of the Air Transport Licensing Board for 1962, para. 8(c).

be attributed to it (in this case including aircraft costs and standing costs of providing the service) then it should ultimately be withdrawn. It has been suggested[1] that B.E.A. has reached the size at which airline average costs become approximately constant so that a small reduction in total size would not drastically affect the average cost of remaining services. Hence in the long run the domestic services ought, if competition under conditions of fair comparability is to be the regulator of our transport sector, either stand on their own feet, or, like the railway branch lines, be abandoned.

A.T.L.B. has explicitly washed its hands of matters of high policy by refusing to consider the effects of its decisions on competing agencies, and by refusing to take account of B.E.A. unremunerative 'social service' flights when considering licence applications.[2] The Board agreed with the Select Committee on Nationalized Industries that the right way of providing for these services is for B.E.A. to receive a Government grant.[3] But B.E.A., like the B.T.C., has traditionally subsidized its weaker services. Unlike the B.T.C., it has not yet been ordered to divest itself of such burdens and has shown no signs of relinquishing them voluntarily. Clearly the Minister must resolve this peculiar situation by a directive either to B.E.A. or to A.T.L.B. if fair comparability is to be ensured between rail and air and between different air operators.

Another interesting anomaly may arise where a combined operation is proposed, requiring both air and road service licences. Both A.T.L.B. and the Traffic Commissioners could require proof that the complementary licence was available before making their own decision, thus causing a stalemate. Or, a more likely problem, one authority may impose conditions on the licence which it grants contrary to those which the other authority, in its deliberations, has assumed to be forthcoming. There is not even, at present, an established sequence for the submission of road and air licence applications for the prospective combined tour operator. Clearly, some formal arrangement to simplify this procedure would be of advantage both to the operator and to the authorities themselves.[4]

[1] S. Wheatcroft, *The Economics of European Air Transport*, Chapter 3.

[2] Report of the Air Transport Licensing Board for 1962, para. 8(e).

[3] Select Committee on Nationalized Industries: Report, 1959, para. 117; Report, February 1962, paras. 51, 53.

[4] See S. MacArthur, 'Tour Operators Should Know Air Licensing', *Bus and Coach*, March 1963.

THE TRANSPORT ACT, 1962

The Transport Act, 1962, is, to quote the preamble to the Act, 'An Act to provide for the reorganization of the nationalized transport undertakings. . . .' It abolishes the British Transport Commission and replaces it by four Boards (Railways, London Transport, Docks and Waterways), and a holding company to be responsible for the large number of companies in which the B.T.C. had a complete or partial interest. This group includes the B.R.S. road haulage companies, the bus and coach companies, Thos. Cook & Son travel agency, David MacBrayne and the Associated Humber Lines shipping concerns and two coach-building companies formerly owned by the B.T.C.

Nothing very revolutionary is envisaged for the organization of the railways. Six regional boards are set up corresponding to the existing areas to share the responsibility for operating the system. The extent to which responsibility is delegated to the regions is to be determined by the Railways Board in conjunction with the Minister. The only really new element in the structure is that the railway system is recognized as a full time responsibility in itself and does not have other major responsibilities appended to it.

The financial duty laid on each of the Boards is similar to that laid previously on the B.T.C., namely, 'to secure that their revenue is not less than sufficient for the meeting of charges properly chargeable to revenue, taking one year with another'. The holding company is similarly required to act in the same way as any normal commercial enterprise. The financial separation of the agencies makes it no longer possible to cross-subsidize consciously within the sphere of nationalized transport.

In order to meet its financial duties the Railways Board is given almost complete commercial freedom. It is relieved of the remnants of the burden of restrictions originally imposed as a control of monopoly power. Only two specific restrictions remain. Firstly, where public passenger travel still remains a monopoly (in the case of London Transport), or is deemed to require control on special grounds (the carriage of police, armed forces and mail), it is left within the jurisdiction of the Transport Tribunal. Secondly, the Railways Board is subject to ministerial direction if it appears to discriminate aggressively and unfairly against coastal shipping where, in real terms, this is the least costly form of transport. Strangely, the Railways Board may only be prevented from such

[1] Transport Act, 1962, Section 18 (1).

aggressive discrimination so long as it is in deficit on revenue account.[1]

The new freedom is not solely restricted to charges. The Board is to be allowed to develop its lands and, if it so desires, to construct and operate pipelines. It is still not allowed, however, to enter into any manufacture or production except for the purposes of either its own business or that of one of the other Boards.

The new legislation spells a major curtailment of the functions of the Transport Tribunal and the Transport Users Consultative Committees. The new Transport Tribunal is to sit in two divisions corresponding to its two main responsibilities;

(a) for London fares and miscellaneous rail charges,
(b) for road haulage appeals.

It only retains power over railway rates and fares in the case of London Transport and in the charges for armed forces, police and mail. In the case of armed forces and police the control is not very important in practice. It is not clear, however, why the railways should be prevented from exercising their powers of discriminatory charging against the G.P.O. any more than against other users. After all, the G.P.O. has the alternative of road transport if this should prove cheaper;[2] if the railways are to pay their way by charging what the traffic will bear then there is no reason why the G.P.O. should not, like other customers, be charged as much as possible towards the railway overheads.

The reduction in the responsibilities of the area T.U.C.C. may appear to be a subtle point. Whereas formerly it could report its approval or disapproval of closure proposals on the basis of any undue hardship involved, it is henceforth only able to report the facts concerning the hardship and suggest measures for its amelioration. Conclusions are left entirely to the Minister who may give or withhold his consent as he sees fit. No provision is made, however, for any payment by the Minister in respect of lines or services which the Board wish to close but over which the Minister does not give his consent. Perhaps this omission is explained by the fact that the Minister did not at the time of the Act envisage any case where he would refuse to give consent. It is certainly possible, however, to imagine cases where it might be considered desirable to retain a service which is not a commercial proposition on economic, social or

---

[1] This peculiar provision is Section 53(2)(b) of the Act.
[2] In June 1963 the G.P.O. began to experiment with trunk road haulage of parcels in East Anglia and expressed their intention to extend the scheme if successful.

strategic grounds. If the railways are to be judged as a normal commercial undertaking then it is surely just that they should be relieved of any loss unwillingly sustained in maintaining such services.

The financial provisions are generous to the Railways Board. The Treasury assumes responsibility for all existing British Transport stock and the 'commencing debt' to the Treasury is apportioned between the new Boards by the Minister. In the case of the Railways Board that part of the commencing debt which, in the opinion of the Minister, is not represented by the written down book value of assets created since 1955, is suspended and does not carry any interest or any liability for repayment at a fixed time. If this does not quite amount to a day of jubilee for the railways when all debts are washed away, it does at least ensure that its current performance should not be hindered by the overvaluation of its inherited assets and a consequent unduly heavy annual interest payment. In addition to this, the liability of the B.T.C. in respect of Exchequer advances to cover its deficits on current account under the Finance Act, 1956, and Transport (Railway Finances) Act, 1957, is extinguished.[1] Finally, it is given five years to get on to its feet and not until the end of this period is the Railways Board required to start breaking even.

The Act is, in general, notably silent on the question of the relationship between various forms of transport. Apart from the provisions to protect coastal shipping it is clearly assumed that free competition will achieve the desired results. Section 55, however, sets up the Nationalized Transport Advisory Council, a body consisting of the chairmen of the Boards and the Holding company and up to seven ministerial appointees, with the duty of 'advising the Minister on questions relating to the co-ordination, or any other aspect, of the nationalized transport undertakings'. It is not clear what are the duties of this Council. By its terms of reference it is confined to nationalized transport and it is presumably unable to advise the Minister on the relationship between rail and road haulage. It could possibly be used to carry out the first pruning of the annual investment budgets submitted by the separate boards in much the same way as the B.T.C. compiled its claim for submission to the Minister. But if the Boards are to act as normal commercial under-

---

[1] The commencing debt of the British Railways Board amounts to £1,900 million. It consists of £900 million ordinary interest bearing debt. £300 million owing to superannuation funds and savings banks and a suspended debt of £700 million. A sum of £500 million, representing the accumulated deficits of the B.T C., was written off before calculating the Railways Board commencing debt.

takings this simply means ranking the schemes in order of expected profitability (with possibly some allowance for differing degrees of risk). Hence, if the competitive process is to be used as the method of co-ordination there does not remain a great deal for the advisory council to advise upon.

THE PIPELINE ACT, 1962

In direct contrast to the Civil Aviation and Transport Acts, the Pipeline Act is designed specifically to protect against competition. The economies of scale in pipeline operation are taken to indicate that this industry is a natural monopoly. Hence the act secures operators against the dangers of duplication and excess capacity which would increase unit costs, whilst reserving powers to the Minister which will enable him to prevent exploitation of any privileges granted.

Trunk pipelines may only be built with the consent of the Minister of Power who will, before granting such consent, satisfy himself that the line is of sufficient capacity to meet forseeable demands without duplication. Once authorized, however, construction is facilitated by powers of compulsory purchase, subject only to the decision of a Lands Tribunal on compensation. On the other hand, once the pipeline is in operation the Minister may insist that it undertake common carrier obligations for the traffics for which it was constructed with no discrimination between different users, at rates determined by the Minister. Additional ministerial powers over mode of construction, safety provisions, repair and examina-tion of lines complete a system of control as rigorous as that to which the railways were subject in the last century. In principle then, the Minister is enabled to ensure high capacity, low cost, development of pipeline techniques without the discrimination and exploitation inherent in the monopoly situation. In practice, much will depend on the extent to which the Ministry is enlightened and well in-formed both about current techniques and about future demand for the pipeline facilities. Indeed the powers vested in the Minister would be best utilized if the sort of investigation of the national industrial structure instituted by the National Economic Develop-ment Council could be used for obtaining well-founded predictions of future requirements.

THE 'NEW LOOK'

The first major practical consequence of the legislation was the Beeching plan. The reshaped railway system which the plan envisaged is described elsewhere (Chapter 10). The major changes proposed are threefold:

(i) Withdrawal of unremunerative passenger services, mostly in rural areas but also on some commuter and holiday routes.
(ii) A similar reduction in the number of depots for freight.
(iii) Concentration on larger bulk and heavier density working, permitting some existing traffic to be transferred to road whilst envisaging the capture of a similar amount of new traffic for the improved trunk freight services.

Such changes are supported on the grounds that, given the freedom to relate charges to cost, the interests of the community are best served by the railways adopting a purely commercial outlook, providing only those services which the public themselves indicate to be necessary by their willingness to meet the costs involved.

This analysis has not gone unchallenged. We shall now consider the major criticisms.

(i) *Costings technique*.[1] The example used in the report to demonstrate the unfavourable characteristics of passenger branch line operations assesses movement costs of a diesel multiple unit as between 4s 0d and 6s 0d per mile and the savings of track and terminal costs as £3,000 per mile where complete closure is involved and £1,750 per mile where only the withdrawal of passenger services is contemplated. But in the actual case studies costed in the appendix to the report the *average* movement costs on lines solely using D.M.U.s is less than 4s 0d, with one case as low as 3s 1d per mile and the *average* terminal, track and signalling savings only a little over £1,000 per mile, again with some extremely low savings (down to £250 per mile). Hence the example discussed is far more unfavourable than many actual cases may be.

In so far as actual costings, and not standard figures, will be used in supporting individual closures the biased example in the report is unimportant. It is clear from the case studies that in certain circumstances relatively low movement and track costs are available. Hence it would appear desirable that closures be considered in the light of the cheapest possible form of operation, examining, for

[1] See D. L. Munby, 'The Reshaping of British Railways, *Journal of Industrial Eco omics*, Vol. 11, No. 3, July 1963.

instance, the possibility of diesel railbus operation between unmanned halts on track maintained only to the standard necessary for the safety of such light operation.

Similarly on the question of 'thinning' services it is assumed far too easily in the report that no solution can lie in this direction. Such a conclusion rests on the relatively greater incidence of track costs as utilization declines. Certainly, for a line with given fixed and movement costs, a greater average number of passengers per train will be required as the number of trains per day is decreased. But we might expect trains at certain times of the day to be better loaded than at others, so that a reduced service may yield a higher average utilization of the trains run. In certain circumstances it thus becomes of crucial importance to show *how many* more passengers per train are required to break even with a reduced service. This will depend on the magnitude of track and station costs. The lower the escapable track costs the greater the likelihood that by thinning services a sufficiently high loading can be obtained to meet the total overhead, as is shown in Table 28.

TABLE 28

Number of passengers per train required for break even with movement costs at 4s 6d per train mile

Trains per day	0 (Movement cost only)	Escapable track and station costs (£ per mile per annum)			
		500	1,000	1,750	3,000
32	27	32	37	45	57
16	27	37	48	63	90
12	27	41	55	76	111
8	27	48	68	99	150
4	27	68	109	171	273

Thus, for instance, a route unable to maintain an average of thirty-two passengers per train for an hourly service might well be able to achieve an average of forty-eight per train for a service of one quarter that frequency. Although 7,168 passengers would be required per week to support an hourly service on this route only 2,688 passengers might suffice to support eight trains a day.

We do not mean to suggest that the Railways Board are incapable of making this type of simple calculation. But they may not be inclined to do so.[1] Dr Beeching's appointment was for five years.

[1] Despite the following disclaimer—'it must be clearly stated that the proposals now made are not directed towards achieving the result by the simple and unsatisfactory method of rejecting all those parts of the system which do not pay already, or cannot be made to pay easily'. *The Reshaping of British Railways*, *op. cit.*, p. 2.

The Transport Act gives the Board a similar period before it is required to meet the financial test of solvency. Reorganization of passenger services might be a very time-consuming procedure with little profit to show for success. As these services are quantitatively so insignificant in terms of total receipts in comparison with freight traffic and main line passenger services, and as the railways planning staff will in any case be fully employed during the quinquennium, it may be tempting to cut the losses in the most convenient way, by closure. Such a built-in bias towards consideration of a relatively short time span would be quite understandable. Unfortunately, the framework of control may be quite unable to detect it. The T.U.C.C.s no longer have the power to assess the validity of arguments for closure. They can merely suggest measures for the amelioration of hardship.[1] Instead the Minister becomes the sole guardian against mistaken closures. In debating the proposals, the Minister of Transport, Mr Marples, stated: 'Responsibility for a decision to close anything where passengers are concerned, be it station or railway, is not the Railway Board's or Dr Beeching's; it rests squarely and fairly on the Government.' In the circumstances this is little consolation, for both the Ministry of Transport and the Government in power had shown no foresight (though a great deal of hindsight), when dealing with questions of principle involved in the modernization plan.

(ii) *The real cost of urban congestion*. As we have shown earlier, an uneconomic allocation of traffic will occur if the price charged for use of road facilities is less than the marginal social cost of use. Thus, in some cases, the community might be well advised to pay the railways a subsidy to carry commuters or freight traffic in urban areas rather than sustain the increased road congestion consequent on the closure of a railway line.

The problem might be alleviated if the use of congested roads was made more expensive. But even a system of pricing for roads related to marginal social cost might not eliminate all scope for error because of the problem of indivisibilities. Consider competing road and rail facilities between two locations. If sufficient traffic travelled by road to make the railway unremunerative, closure would be expected to ensue. Transfer of rail traffic to the existing

---

[1] In any case these bodies were never sufficiently expert or well informed to argue very seriously with the B.T.C. prior to 1962. Attempts to question the B.T.C. submission on question of fact were even refused a hearing by the T.U.C.C.s on occasion. See G. Mills and M. Howe, *Journal of Public Administration, op. cit.*

road facility might then cause the marginal social cost of road transport between the locations to rise substantially. Consequently the total real cost of transport between the two locations might be increased, despite the existence of prices related to marginal cost for both road and rail, due to the existence of indivisibilities.

Even if the introduction of a more sensible pricing system for urban road use was not deemed practicable it would still be prudent to secure the application of cost-benefit analysis to the various alternatives before any change was sanctioned. Indeed, on May 2nd the Minister specified certain cities where cost-benefit analysis would be desirable and threatened to reject any closure proposals, made separately, which he considered to be rightly part of a major conurbation problem.

It is difficult to assess the quantitative significance of the argument. Compared with the 879,000 estimated to commute daily by rail in London, the 40,000 commuting to Glasgow and 34,000 to Manchester do not appear to be large flows.[1] In both of these latter conurbations, however, rail transport accounts for over 25 per cent of the daily flow and yet many local services are scheduled for closure. The crucial question is what proportion will transfer to bus travel and what proportion to private cars. If a large proportion were to transfer to the former the problem of congestion might not be increased too seriously; if to the latter, the pressure on road facilities might only be met at enormous expense.

In the text of the Beeching report the potential validity of this argument is acknowledged thus: 'The right solution is most likely to be found by "Total Social Benefit Studies" of the kind that are now being explored by the Ministry of Transport and British Railways jointly.'[2] Nevertheless, Appendix 2 to the Report contains amongst the passenger services to be withdrawn many which are discernibly commuter services, without any distinction of type being made.

(iii) *Hardship in rural areas.* The third area of challenge has concerned the hardships involved for a relatively small number of people in rural areas. It is argued that not all those deprived of their rail service are able to provide private means of transport. Nor does

[1] National Institute, *Economic Review*, May 1963, pp. 23 ff.
[2] 'The Reshaping of British Railways,' p. 22. Again on p. 56 ' . . . in the case of suburban services around some of the larger cities is there clear likelihood that a purely commercial decision within the existing framework of judgment would conflict with a decision based upon total social benefit. Therefore, in those instances, no firm proposals have been made but attention has been drawn to the necessity for study and decision.'

it appear that the existing structure of control of the bus industry is capable of ensuring a continued service in all areas. But great care must be exercised in drawing conclusions from these premisses. Neither a subsidized railway nor a subsidized bus service is necessarily the most economical way of providing what is deemed to be the basic transport requirement of such areas. For instance, an abrogation of the P.S.V. and road service licence requirements in such areas, to permit local tradesmen or even car owners to carry passengers for reward, might suffice. Both the possibilities and requirements will vary from area to area. To arrive at the most satisfactory and economical arrangement where a bus service of the traditional type cannot be remuneratively operated requires both detailed local knowledge and full appreciation of the general advantages and disadvantages of various alternatives. Under the present régime local needs are considered separately by the T.U.C.C.s in vetting closure proposals and the Traffic Commissioners in dealing with licence applications. Though both of these bodies are in frequent contact with the local authorities concerned, in both cases their task is limited to a decision concerning *one* form of transport, without investigation of the full range of alternative methods of meeting local needs. A far more thorough approach might be obtained by a local Government agency assisted by a national advisory service. The financial problem of how local services were to be subsidized could then be treated separately, although a certain degree of local financial responsibility would encourage a more economical use of resources.

(iv) *Development policy.* The availability of adequate transport facilities is an important factor in determining the location of industry. As part of a positive location policy it might thus be desirable to provide facilities in advance to stimulate development. Or, on the other hand, it might be sensible to temporarily subsidize an existing location on the grounds that a changed location of industry might necessitate an entirely new investment in social facilities (housing etc.) whilst such facilities in the old established areas were under-utilized. The arguments need to be handled very carefully however. In the first place 'adequate transport facilities' does not necessarily entail local rail depots and stations. For instance, in planning for the north-east, private sidings for any customers large enough to justify the capital expenditure involved, and one liner train depot each on Tyneside (Newcastle) and Tees-side (Stockton), might suffice to give the whole area an efficient road-rail link with the rest of the country.

Secondly, in any area social capital needs to be replaced or extended from time to time. It would be uneconomic to extend social facilities to serve an industrial complex located as a result of transport or other subsidies unless the area promised to stand unsubsidized at some time in the future. Development area policy not based on genuine low cost production potential may serve to ossify a distribution of population which is uneconomic both by region and by industry. Where the views of the Government and of the Railways Board of the future potential of an area conflict the Minister of Transport has the power to reject any relevant closure proposal and enforce a subsidy. Unfortunately there is currently no convention whereby the cost of the subsidy can be transferred from the Railways Board accounts to the appropriate public account. Regional development policies may justify the provision of subsidized rail services but where it is enforced on the railways in conflict with their normal commercial duties it seems both illogical and unjust to expect the Board to meet the cost. Even if the railways are capable of carrying this subsidy and meeting it through their powers of discriminatory pricing on other traffics it constitutes a distribution of a social burden on the basis of the elasticity of demand for transport services, a peculiarly hidden form of taxation. If transport charges are based effectively on 'what the traffic will bear' this may not distort the allocation of traffic between alternative transport agencies but it will raise the delivered cost of traffics for which rail transport has a significant cost advantage, notably coal and heavy minerals.

The arguments which we have adduced hold for the tourist industry as for any other. If it is deemed desirable to sustain viable access to a particular resort and the seasonal flow of traffic can be carried most cheaply by retaining an unremunerative rail route then the financial responsibility should rest either on the Ministry or on relevant local authority.

(v) Employment

The sternest and the most publicized immediate challenge to the plan came from the National Union of Railwaymen on the employment issue. Though the envisaged staff reductions in the first five years of the plan were in total less than the realized reductions in the previous five, and the proposed redundancy and disturbance provisions generous, treatment of the unions was most tactless. The statistics on redundancy contained in the report were meagre and hardly to the point in that they failed to give any adequate breakdown of expected redundancies and vacancies by region or by occupation and grade. Admittedly the necessity for dismissals does depend to a considerable extent on the pattern of natural wastage but, in view of the degree of speculation in the Press and the obvious concern of employees, this deficiency was almost certain to be subject to the most pessimistic and alarmist interpretations. Under threat of strike action in early May the Board made the position a little more clear. The fact that union pressure was directed towards securing clarification rather than major concessions suggests that with a little more effort on the Board's part the proposals could have secured a far less grudging acceptance from the unions than in fact transpired.

As a purely economic question railway redundancy only involves social cost in so far as alternative employment is not forthcoming or is less productive in real terms than rail employment. Given an adequate demand for labour in the economy as a whole, the net increase in unemployment as a direct result of the plan should not at any time exceed 2,000 according to N.I.E.S.R. estimates.[1] Because of the very low net output of the railway workers laid off it is very unlikely that they would be re-employed in any concern where they would achieve a lower output in value added terms. Thus, whilst recognizing the desirability of alleviation of individual hardships involved we suggest that the economic losses as a result of unemployment caused by the proposals are not likely to be of any great significance.

The value of Beeching

The Beeching plan has been widely acclaimed as a thorough, but overdue, application of commercial principles to railway operation. Certainly many of the savings which will result will be real as well as financial and it would be most unwise to attempt to delay the implementation of the unchallenged recommendations simply because there is doubt in some other cases whether the public interest would be best served in the proposed way.

Not least of the virtues of the report is that it has challenged its critics to produce a much more thorough and rigorous working out of the real cost elements in the transport problem than has hitherto been attempted.

THE LABOUR PARTY'S APPROACH

One feasible prospect for the 'sixties is that of a Labour Government. Therefore, if there is a specifically Labour policy for transport, it deserves mention in this chapter. The Party has not published any

[1] *National Institute Economic Review*, May 1963.

official policy document on transport for some years now. Consequently we must infer its current attitude from a number of separate and sometimes conflicting sources. Two main features distinguish the Labour approach from that currently applied, its attitude to the nationalized sector and its attitude to regional planning.

In its policy document 'Signpost to the 'Sixties' the Party states its intention to remove any remaining impediments which might prevent the nationalized concerns from undertaking whatever operations they consider advantageous. It further states: 'Where competition creates not efficiency but chaos in a key sector of the economy, there an expansion of public ownership may be necessary to put this right. That is our case for creating an integrated and publicly-owned transport sector.' This might be interpreted simply as a reiteration of the established intention to renationalize road haulage. There does appear to be some question as to whether this policy should be retained however. One element has recently come out publicly in opposition to renationalization[1] and in his speech in debate on the Beeching plan, Mr Harold Wilson spoke not of renationalization but of removing any restrictions on the expansion of B.R.S. So perhaps the Party is in the process of changing its mind. Public ownership of transport might, however, be extended in one other respect. 'Signpost to the 'Sixties' indicates that wherever public subsidy is necessary there should also be at least partial State control of undertakings concerned. Rural passenger transport might thus be subject to some extension of public ownership on these grounds.

The second distinguishing characteristic of Labour transport policy would seem to be a disposition to regard it as part of a number of regional problems involving several different agencies rather than as a number of separate industry studies. For instance, in its criticism of the Beeching report the Party concentrated on the need for similar criteria to be applied simultaneously to both road and rail transport in each area. Similarly it has been argued that an essential step towards the efficient use of resources is that all transport investment decisions are made on the same basis and perhaps by the same body, instead of being made entirely independently. To this end it has been suggested that regional and central planning agencies should be responsible for both road and rail investment.[2] In addition, various suggestions have been made for the control of urban road use without, apparently, very much thought having yet been given to the matter.

Thus as far as technique is concerned, nothing very new has

[1] See 'Transport is Everyone's Problem' *Socialist Commentary*, April 1963.
[2] *Socialist Commentary, op. cit.*

been suggested, though the more thorough regional investigations which are advocated might involve more widespread application of cost benefit studies. The Labour Party might, by nature, be less suspicious of decision procedures involving full social benefits rather than private commercial benefits. Despite the fact that the slogan of 'integration' is still used it is likely that even the Labour Party policy would essentially consist of free consumer choice between competing transport agencies.

CONCLUSIONS

The prospect for the 'sixites is one of co-ordination through competition. To this end the railways have been granted complete commercial freedom and the road haulage sector is under review.

But difficulties remain. Competition will only produce a desirable outcome if private costs are related to social costs; this is not currently the case in urban road use. To permit competing agencies to attempt to make full use of fixed capacity by discriminatory charging may be part of the system; it can result in a wasteful use of resources if competitors continue to provide alternative facilities where the total traffic is not sufficient to support them, as seems to be the case for air and rail transport on some trunk routes. Similar dangers arise where competing agencies do not use comparable investment criteria; this again seems to be relevant to competition between road and rail. Furthermore there are problems of transition where the operations of an agency are being radically changed and of social policy where remunerative transport services cannot be provided.

CHAPTER FOURTEEN

British Transport Policy and the Common Market

Though negotiations for British entry into the European Economic Community have for the present been discontinued the possibility of British entry at some time in the future makes it desirable to consider the implications for the transport sector of such a step.[1] Article 3 (e) of the Treaty of Rome indicates that the activities of the community shall include the inauguration of a common transport policy. Thus British entry would necessitate the assimilation of British policy into a common community policy.

This common policy is envisaged in Articles 74–84 of the Treaty. The Council of Ministers is empowered to lay down common rules applicable to international transport and conditions for the admission of non-resident carriers to other member states, though due account must be taken of the effects of such provision on regional employment conditions (Article 75). There are, however, only three more specific indications in the Treaty of the type of common policy envisaged, namely:

(i) Discriminatory rates or conditions of carriage based on the country of origin or destination of any traffic are to be abolished (Article 79).

(ii) Support of home industry by specially determined rates is to be prohibited except where the explicit consent of the Commission is granted (Article 80).

(iii) The charges for the crossing of national frontiers shall be related to the costs involved (Article 81).

So far, so good; none of these provisions appears to be incompatible with British transport policy. Why then should there be any concern about the outcome? The answer is twofold. Firstly E.E.C. policy is likely to be based, at least to some extent, on the precedents provided by E.C.S.C. transport policy, which does not accord very easily with British policy. Secondly, a Memorandum on Transport Policy produced by the Economic Commission in April 1961, and an Action Programme based on it in May 1962, outline proposals apparently incompatible with recent British policy.

[1] For a fuller discussion see 'Transport in the Common Market', *Planning*, Vol. 29, No. 473, July 1963.

E.C.S.C.

Because of the great quantitative significance of coal and steel traffics in Western Europe the policy adopted by E.C.S.C. is bound to exert a strong influence on subsequent E.E.C. policy. In particular the fact that 70 per cent of E.C.S.C. traffics are carried by rail makes the policy towards rail freight most significant.

Discrimination and the publication of rates

The Treaty of Paris, which set up E.C.S.C., prohibited discriminatory transport rates in general and discriminatory rail rates based on national considerations in particular. The crucial point, however, is that in order to eliminate any discrimination it was laid down that all charges and conditions of transport be either published or notified to the High Authority. Thus it was the explicit aim of the Treaty to establish a system of published rates and conditions similar to that which British transport policy has so recently decided to destroy.

When considering British policy we have argued that the desire for greater flexibility of charges is based on respectable economic principles. Does this mean that E.C.S.C. was not only perverse but also positively 'uneconomic' in embracing a contrary policy? We can only answer this question by spelling out the E.C.S.C. transport situation in a little detail. As, apart from waterborne traffic on the Rhine, most of the international traffic is carried by rail the main difficulties with which the High Authority had to contend were those arising out of discriminatory railway practices.[1]

There were four main features of the international rail transport organization of Western Europe which conflicted with the declared aim of the Treaty of Paris 'to assure to all consumers in comparable positions within the common market equal access to the sources of production'.

(i) *Conscious national discrimination.* The railway companies concerned are State owned monopolies. As such most of them have been consciously used to support national industry by discrimination in favour of those traffics contributing to the prosperity of the national heavy industries and against those competing with them. Transport subsidies were also used, particularly in Germany, to encourage the decentralization of industry for political reasons. Such measures distorted the pattern of traffic to favour internal rather than

---

[1] Similar difficulties arose with water and road traffic, see J. E. Meade, H. H. Liesner and S. J. Wells, *Case Studies in European Economic Union*, pp. 336–405, for a fuller description of E.C.S.C. transport policy.

international trade, often increasing the real total cost of transportation. However, because the three main producing countries, Belgium, France and Germany, were all pursuing similar policies, they cancelled each other out to a very large extent so that no significant national advantages were achieved by the system. Consequently the suggestion of the High Authority that such discrimination be abolished was readily accepted. In abolishing discrimination rates were permitted to be equalized either downwards or upwards so that the community was not committed by this step to a common level of charges.

(ii) *Broken rates.* In the community countries rail rates consist of two parts, a fixed terminal fee and a transport charge proper, which, though varying with distance, is tapered in favour of the longer journeys. International traffic was formerly treated as though undertaking separate journeys in each of the countries traversed. This made international transport rates higher mile for mile than those for internal traffics for two reasons. Firstly, there were two or more terminal charges to be paid even though transit was unbroken. Secondly, the total mileage would be split into shorter component journeys and the higher rates per mile applicable to the shorter journeys charged on each of the national railways involved. The costs of crossing frontiers were hence similar to the imposition of a general tariff and as such were incompatible with the aims of E.C.S.C.

The Transport Commission of the Community devised the following solution. Only one terminal charge was to be levied, that being the average of the charges in the separate countries involved. For the carriage charge the journey was to be treated as a single mileage and a weighted average taken of the rates applicable if such a journey length had been completed internally in the individual countries. The weights applied in averaging these rates were to be in proportion to the mileage run in the various countries by the particular traffic concerned. Again, however, there was no commitment to a common level of charges.

(iii) *The taper.* We have already considered the effect of tapering rates when charges were broken at national frontiers. Even when this difficulty was abolished the continuation of different tapers or 'rates of degression' could still distort relative market prices. The steeply tapered rates operated in Germany, for instance, secured more favourable treatment for distant locations in Germany than for similarly remote locations in other countries. Even E.C.S.C. was unable to secure any satisfactory equalization of the rates of degression except for the shorter journeys where the differences had not been important in any case.

(iv) *The general level of charges*. Because the railway companies are State owned monopolies it is possible for them to continue to operate indefinitely with an unprofitable rates structure. Distortions are likely to result. To take an extreme example, where only one national railway was offering services below cost such rates would automatically improve the competitive standing of that country's exports and, depending on the location of the points of production and consumption, might affect the competitive position of home products with respect to imported alternatives either favourably or unfavourably. Similarly, an unduly favourable classification of an individual commodity might result in an uneconomic location of industry within the common market.

In so far as costs of transport differ between countries, there are grounds for different rates to be charged, so that the formulation of any precise rules for the harmonization of rates is problematical. To insist that rates be equal to direct cost, however adequately defined, is to destine railways to incur losses. To allocate overheads according to some arbitrarily selected schedule is to risk losing traffic to competitive agencies and hence to make inefficient use of existing capacity. The British answer to this dilemma is to allow the railway company to recoup its overheads as best it can by charging 'what the traffic will bear'. It is assumed that gross exploitation of this freedom is prevented by competition from other agencies. But in the case of E.C.S.C. international road transport capacity was severely limited. There was thus considerable scope for undesirable discrimination and the High Authority, in view of the unsatisfactory records of the national railway in this respect, insisted on publication of rates as the only sure way to prevent its occurrence.

Special rates
Despite the publication of rates in both Germany and France before the establishment of E.C.S.C., substantial proportions of heavy mineral traffics were carried at special rates. These were not always the result of conscious national discrimination or regional support policies but were also granted where:

 (i) They were necessary to obtain traffic in the face of competition.
(ii) They reflected specially favourable costs for particular traffics or routes.

Both of these cases were accepted in principle by E.C.S.C. as conditions which might justify special rates. The High Authority suggested, however, that favourable cost conditions could be incorporated in the published rates.[1] Concerning competition it considered that rail transport was sufficiently superior in most circumstances for special rates to be an infrequent exception which should only be granted after detailed consideration of the case concerned. The High Authority thus assumed that the general rate structure would not be seriously affected by permitting these exceptions in principle.

Publication of rates and road transport

Application of this principle of publication of rates to all agencies could be a costly mistake. The argument for it rests on the great scope for uneconomic discrimination which the railways in particular cannot be trusted to exercise in secrecy. But this scope is itself the result of three particular characteristics of railway operations, namely:

(i) The high proportion of joint overhead costs which cannot be objectively allocated between traffics.

(ii) Monopoly operation.

(iii) The possibility of deficits for a nationalized railway undertaking.

None of these considerations is directly relevant to road transport. Operators are unable to obtain direct subsidy in the case of unprofitable operation, and, because of the smaller operating unit, overheads may more easily be attributed to particular traffics. Hence the haulier, out of self-interest, will be unlikely to carry traffics below cost except by accident, and, because of competition from other hauliers, will be unable to recoup a disproportionate part of the unallocated overheads from 'foreign' traffic. Hence the publication of road rates is an unnecessary precaution against undesirable discrimination.

Not only is it unnecessary however; it is positively harmful. Road services may be costed with more precision than rail. The establishment of mileage scales, however complex, might relate only roughly to the costs involved in the carriage of particular consignments. If the published charge for a class was set fairly low, then, because of the wide spread of the costs of carriage in reality, much of the traffic offered to the haulier in that class would cost more to carry than the permitted charge. There would thus be a tendency to evade such

---

[1] For instance, the French railways already operates a published tariff containing most complex provisions to adjust charges according to the volume of traffic handled at the stations used.

traffic, to force it to be carried by other means. In particular any operator unfortunate enough to be subject to common carrier obligations would have a great deal of such unremunerative traffic thrust upon him. If, on the contrary, rates were set high to avoid the acceptance of unremunerative traffic there would then be traffic which could be carried at a cost considerably below the published price. If the published price was effectively enforced many entrepreneurs might be tempted to provide transport on own account in these circumstances and traffic would be transferred from public to private haulage irrespective of the relative costs of the two agencies in real terms. The pressure for the public haulier to undercut the published rate would be very strong. With such a proliferation of small operators in the industry enforcement might prove an impossible task. The greater ease of evasion for the small operator, both evasion of traffic where cost was above the published rate and evasion of the rate when cost was below it, would place him in a favourable position relative to the large firm irrespective of the efficiency and real costs of the different scales of operation.[1]

Publication of rates will therefore militate in favour of private and against public road haulage, in favour of small as opposed to large scale operation and in favour of the unregulated as opposed to the common carrier. In each case the advantage would not necessarily be based on the real cost of carriage.

E.E.C. POLICY

In April 1961 the Commission of E.E.C., at the request of the Council of Ministers, submitted to them a memorandum, known as the Schaus Memorandum, setting out the principles of a common transport policy. In May 1962 this was spelt out in the form of an Action Programme including some suggestions of detail and timing. The proposals outline three main objectives of a common policy:

(1) The removal of any obstacles which transport might represent on the establishment of the general common market.
(2) The creation of healthy competition of the widest possible scope.
(3) The development of a transport system which will prove a powerful stimulant for the growth of trade and widening of markets.

In order to meet these objectives a series of principles are enunciated; operators should be financially autonomous, subject to equal

[1] See D. L. Munby, 'Fallacies in the Community's Transport Policy', *Journal of Common Market Studies*, Vol. 1, No. 1, May 1962.

treatment by the authorities and have a very large degree of commercial freedom, users should have free choice of agency and investments in transport through the community should be co-ordinated.

The proposals differ in several important respects from the structure operated for E.C.S.C. traffics.

Subsidy

One of the basic principles recommended is that transport enterprises should be financially autonomous. Whereas E.C.S.C., as a restricted body, was unable to deal with the problem of the general level of charges the proposals argue that national transport systems within E.E.C. should not normally be allowed to offer subsidized services, even where the subsidy was applied without conscious national discrimination. This implies that enterprises, however owned, should be able to balance receipts and expenditure from their own resources and should have free choice of the means to achieve the balance. It is made clear that road operators should bear a share of the costs of the roads infrastructure. As this stands it accords very closely with the attitude to the nationalized transport undertakings in Britain at present.

Competition

There are several suggestions designed to increase the extent and effectiveness of competition in transport. For instance the general provisions concerning the freedom of establishment and the regulation of monopoly are applicable to transport as to other sectors.[1] Non-resident carriers should be allowed into the national market on the same terms as residents, a suggestion which would seem to require the abolition of flag privileges in the case of inland shipping and a more flexible system of licensing or authorizations in the case of road transport. The Commission would like to see the present system of bilateral quotas for international road transport replaced by a system of Community licences which would enable hauliers to organize more effectively for international traffic.[2]

The publication of rates

Whereas E.C.S.C. advocated the publication of a fixed schedule of charges the proposals for E.E.C. policy appear to be more liberal,

---

[1] In contrast, however, in November 1962 the Council of Ministers passed a regulation suspending the application of the Cartel Regulation to inland transport for three years, during which period special regulations should be devised for transport.

[2] *Planning, op. cit.,* pp. 270–71.

involving the publication only of maximum and minimum charges for each form of transport within which operators are to be permitted to charge as they please. The limits in this system of 'tarification a fourchettes' are to be worked out by public authorities and the professional organizations.

Now the actual cost per mile of carrying a particular physical commodity will vary tremendously according to the nature of the route, the total loading obtainable, the obtainability of returned loads and so forth. Therefore, unless the limits are set very wide there will be some traffic forthcoming whose costs are outside the limits, either on the upper or the lower side. Either losses will be incurred on some traffic or, at the other extreme, traffic which could have been carried at little cost will be lost due to the inability of the operator to quote a low price.[1] In the case of general merchandise the spread of costs may be even greater than for E.C.S.C. products, due to the uncertainty of loadings, with a resulting accentuation of the defects of fixed charges. Such a system is weighted against public carriage in general, large scale operation in particular and above all against operators saddled with common carrier obligations.

Our argument against 'tarification a fourchettes' has so far been against narrow limits. This is justified by the attitude of the Commission which envisages in rail transport a range between limits of only 5 per cent of the total charge, while the suggested limits for road haulage in 1966 are from 10 per cent below to 30 per cent above the average for each class. Moreover the Commission argues that increased competition may be expected to reduce the range even further. The validity of this argument is doubtful. On the contrary, as the disadvantages of the system become obvious to operators, it will be in their interests to secure reductions in the lower limiting price (to enable them to compete for any traffic which they could profitably carry) and to increase the upper limit (to avoid carrying unprofitable traffic), hence increasing the range of the fork.

In view of these serious disadvantages we might well ask why the Commission is so enamoured with publication of rates as the basis of the common transport policy. There are three possible reasons.

Firstly, in so far as the railways retain an effective monopoly of certain traffics the Commission may not wish to trust the national railway systems with powers of discrimination lest they use them in the traditional nationalist manner. Even when each railway system is required to be financially self-sufficient they may retain sufficient scope, by abstaining from profit maximization, to practice this form of

[1] To transport on own account if not to other public transport agencies.

R

discrimination. In a way this is a thoroughly bad reason for publication as it is based on the sort of mutual national mistrust which E.E.C. must eliminate if it is to be successful. Thus this type of justification is, at best, only acceptable for an interim period of readjustment.

A second argument may also be based on mistrust. One necessary condition for freedom of charging to lead to a sensible allocation of traffic is that operators will not charge below direct costs. In fact they may do so not only because they consciously wish to subsidize but also because they do not know what the escapable costs of a particular service are and in this state of uncertainty feel that the safest course is to quote whatever rates are necessary to retain their traditional traffics. If railway managements, whose specific task it is to obtain information and exercise judgment on such questions, cannot be trusted to avoid mistakes, then there is more rather than less reason to suspect outside control of charges policy. As for road haulage, ruinous competition based on ignorance, either of costs, in charging, or of prospects, in determining capacity, is unlikely to arise during times of prosperity. In any case limitations on capacity and the high technical qualifications to be required of hauliers should eliminate the danger.

A more genuine reason for publication of rates may spring from the implications of charging 'what the traffic will bear'. The upper limit of such a charge is set either by the cost of alternative forms of transport or by the complete exhaustion of the advantages of buying the service to the consignor. Thus, if no alternative transport was available, the upper limit would only be reached when the delivered price of the traffic offered had been raised as near as possible to the delivered price of competing goods from other sources or produced locally. Now although it would not pay the railways to quote such a high price that the traffic was lost, it might be possible for the railways to extract from the producer most of the difference in production cost resulting from efficiency. Thus incentives to productive efficiency, though not eliminated completely, might be reduced in magnitude and hence in effectiveness by charges based on 'what the traffic will bear'.

On the other hand, assuming that there are no artificial barriers to entry into the industry, that road competitors are able to quote prices related to cost and that transport concerns are required to be financially autonomous, a series of market decisions will ensure that total transport requirements are provided at a minimum total cost. Hence this argument for publication of rates may be stated in terms of a choice between the minimization of costs in the short run and the possibility that by reducing incentives future productive efficiency may be hampered. For general merchandise at least competition from road transport could become sufficiently keen as to minimize the deleterious effects of charges based on 'what the traffic will bear'. Publication of rates would then be a very clumsy sledgehammer with which to crack a relatively innocuous nut.

THE SCOPE FOR COMPROMISE

The common transport policy is still largely in the stage of formulation. Britain may justly claim to accept the overall aims and principles of the common transport policy as set out in the Memorandum. It might be argued that monopoly exploitation could be prevented by the creation of an independent tribunal of appeal, similar to our Transport Tribunal, and that ruinous competition could be eliminated, if necessary, by the control of total capacity through a licensing system. Competition, under such a régime, might be practised more freely and with a closer and more sensible relation of prices to costs than under a system of fixed published rates.

The Schaus Memorandum and the Action Plan are still only proposals. The attitudes towards transport in the member countries of E.E.C. differ so radically[1] that the Council of Ministers has been unable to reach agreement on them. It is unlikely, in view of this, that any substantial progress will be achieved until January 1966 when the Council become able to reach decisions by a qualified majority. By that time the question of British participation in the community may be finally resolved and her interest and influence in the promotion of a common transport policy either tremendously increased or totally eliminated.

[1] See A. J. Harrison 'Transport Policy in the European Economic Community', series of articles in *Modern Transport*, January–May 1963.

CHAPTER FIFTEEN

Conclusions

We have been considering the problem of rational provision and use of transport services primarily within the context of a market system, albeit a far from free or perfect market. By virtue of these market imperfections and more particularly because of the indivisibility of many transport costs we have been unable to state an unambiguous pricing rule, such as the Hotelling-Lerner rule, which will automatically and in all circumstances produce a desirable allocation of traffic between agencies. Nor have we been able to simplify the problems of investment into a single criterion capable of universal application.

The problem of co-ordination is thus most complex. The market mechanism has nevertheless been advocated as the basis of a solution despite its imperfections because it offers the most feasible method of securing an acceptable evaluation of the different qualitative facets of the transport product; this is especially important where a large number of users and providers of transport are concerned. The problem thus resolves itself to that of recognizing and handling those situations in which an unmodified market mechanism will cause a wasteful pattern of resource allocation or use. There are three types of situation where this may be the case.

Firstly there is the situation where free competition would lead to a wasteful duplication or over-extension of capacity. Railway construction in the mid-nineteenth century certainly fell into this category; the Beeching plan is largely concerned with the elimination of some of the remaining duplications. Similarly, it was largely on these grounds that both public road haulage and public road passenger transport were regulated in the early nineteen-thirties. In these parts of the transport industry the prospect of waste rested on the existence of many small operators willing to price down to direct cost and to remain in the industry for negligible returns, whilst any bankrupts would be rapidly replaced by new entrants. These are conditions only found in depressed times such as the 'thirties. We may therefore sensibly doubt the validity of this particular justification for the continued control of road haulage at the present day.

Wasteful duplication may also occur as the result of competition between two large operators if both are unduly sanguine concerning their competitive standing in the long run and yet, in the short run, are able to carry the losses involved in competition in an overprovided sector. This could occur if both competitors had monopoly profits from some services with which to cross-subsidize the unremunerative ones, or if they were able to continue in operation for long periods despite overall unprofitability. This seems to be a fair description of the competition between rail and air transport over some routes. There is therefore a strong case for intervention, or at least explicit recognition of the situation, though one must bear in mind the need to exercise caution lest future technological advance is prejudiced by current intervention.

Secondly, the cost of the pricing procedure itself may be too high. This has long been the case with roads, tolls charged directly for use, with a few notable exceptions, being both wasteful and unpopular. More sophisticated pricing methods have recently been advocated, though it has not yet been conclusively demonstrated that the cost of introducing a new system would be less than the value of the benefits to be obtained from it. Where a sensibly based pricing system cannot be operated, an alternative form of investment criterion, independent of prices and profits, is required. The development of cost benefit analysis, which should be accelerated by the proposal of the Ministry of Transport to commission, jointly with the Railways Board, further studies of its application to rail projects, is an attempt to meet this need.

Thirdly, a wasteful pattern of resource allocation may ensue where the social costs of providing a service are in excess of the private costs. This arises particularly in the case of urban road congestion where the marginal charge differs most significantly from the marginal social cost. In principle though, a similar misallocation could result whenever the relationship between marginal social costs and marginal private cost (i.e. price to the consumer) differs between competing agencies. There are two possible solutions to this problem. Either prices can be modified to reflect marginal social cost or profitability must be rejected as a poor indication of social benefit in these cases. Despite some of the Beeching plan proposals, it may be considered economically desirable to support certain unremunerative rail services on these grounds. The Hall report, for instance, when talking of urban congestion, concludes: 'The availability in certain cities of a rail network for internal and suburban passenger movement is the most hopeful feature of the situation. Rail transport in the cities which have it is an asset which should not lightly be eroded.'[1]

[1] *The Transport Needs of Great Britain in the Next Twenty Years*, Ministry of Transport, 1963, para. 71, p. 17.

In addition to the strictly economic reasons already adduced it may not always be deemed desirable to follow the dictates of a market mechanism if questions of income distribution are also at issue. For instance, it may be a matter of public policy to support certain unremunerative public transport services in rural areas because undue hardship would be caused to a few individuals as a result of withdrawal. Or it may be deemed desirable to use a transport subsidy in support of a development district policy. It is not within the competence of either the economist or the transport operator to question the value judgments involved in a decision to subsidize. It may, however, be perfectly legitimate for them to indicate exactly what value can be obtained for money in subsidies to different forms of service. For instance, a decision to subsidize rural transport should not necessarily, without question, be taken to imply support of the traditional forms of transport in that area.

A major part of the problem at this stage is that of securing adequate information on which to base decisions. A great deal of useful data has already been provided by such investigations as the London Traffic Survey and the Ministry Road Goods Transport Surveys as well as by the Road Research Laboratory. The immediate problems which arise are being considered by an Inter-departmental Working Party on Traffic, Highways and Urban Development whilst overall problems of city planning in the longer term, involving the questions of environment as well as those of communications, have recently been considered by a group under the leadership of Mr C. D. Buchanan. Nevertheless there is still scope for further thought and research in several fields, of which the Hall Committee have suggested three.[1]

(a) Development of improved criteria for allocation of new investment within each agency, between the agencies, and between the transport sector and the rest of the economy.
(b) Study of the nature of demand for transport services including its elasticity with respect to income, price and quality.
(c) Study of the costs incurred by transport operators both directly in terms of expenses and indirectly in terms of congestion and waiting losses.

Above all there may be outstanding problems of public relations and education. A public conditioned to a very prodigal use of cheaply provided public transport facilities may not be willing to face the structural changes necessary if waste is to be avoided, even though it would ultimately benefit from such changes. For instance

[1] *Op. cit.*, para. 74, p. 18.

the argument for a more rational pricing system for roads is both complex and difficult; it would require considerable explanation to make it publicly acceptable.

Similarly it may be necessary to undermine the automatic identi-fication of profitability with efficiency and unprofitability with inefficiency. Such a simple identity may be totally inappropriate in a complicated economic situation involving huge monopolies and Government intervention, for various reasons, on a similarly large scale. Prices and profits must, of course, retain a very important place in the system, but their role should be treated with real under-standing rather than mystical acceptance. What is required, and what this book has attempted to produce, is not simply a fuller knowledge of the facts of transport operation but also a more realistic applica-tion of economic principles to the complications of the transport sector.

INDEX

For Product Safety Concerns and Information please contact our EU
representative GPSR@taylorandfrancis.com
Taylor & Francis Verlag GmbH, Kaufingerstraße 24, 80331 München, Germany

9 780367 740771